THE FUTURE AIR NAVIGATION SYSTEM (FANS)

For my son Jeffrey, I am always here.

The Future Air Navigation System (FANS)

Communication Navigation Surveillance Air Traffic Management

VINCENT P. GALOTTI Jr
International Civil Aviation Organization

Ashgate

Aldershot • Brookfield USA • Singapore • Sydney

Published by
Ashgate Publishing Company Limited
Gower House
Croft Road
Aldershot
Hants GU11 3HR
England

Ashgate Publishing Company
Old Post Road
Brookfield
Vermont 05036
USA

Reprinted 1999

British Library Cataloguing in Publication Data
Galotti, Vincent P.
The future air navigation systems : communication navigation surveillance air traffic management
1.Navigation (Aeronautics) 2. Navigation (Aeronautics) - Problems, exercises, etc.
I. Title
629.1´3251

Library of Congress Cataloging-in-Publication-Data
Library of Congress Catalog Card Number: 96-85967

ISBN 0 291 39833 2

Printed in United Kingdom at the University Press, Cambridge

Contents

Figures

Foreword

Philippe Rochat, Secretary General
International Civil Aviation Organization (ICAO)

The International Civil Aviation Organization, in cooperation with the civil aviation community, has embarked upon a formidable challenge: the implementation of a global air navigation system to serve every nation on the face of the earth. At the heart of this unique undertaking, is the visionary concept of Communication, Navigation, Surveillance/Air Traffic Management, better known as the CNS/ATM Systems Concept. This is the most far-reaching project ever initiated in the history of civil aviation and, as such, it will require an unprecedented level of collaboration by all government agencies involved in air transport and all members of the airline industry. It is therefore crucial that every player involved in the process, gain an intimate knowledge of the concept, its advantages and its requirements.

In his momentous work, *the Future Air Navigation System,* Mr. Vincent Galotti, a technical officer in ICAO's Air Navigation Bureau, sets us on the right path with a historical record and detailed description of the CNS/ATM Systems Concept, in a useful and easy-to-understand format. I am proud that ICAO was able to support Mr. Galotti in this endeavour, and I heartily recommend this book as a practical and valuable guide to all those with an interest in maintaining safety, reliability and efficiency of our world air transport system.

Philippe Rochat

Foreword

Marvin L. Smith, Ed.D, Director, ATC Program
Embry-Riddle Aeronautical University

We have all read much about future air navigation systems (FANS) and the Communication, Navigation, Surveillance - Air Traffic Management (CNS/ATM) Systems Concept that will bring about the much needed improvements in aviation system safety and capacity. Unfortunately, up until now, there has been a general lack of understanding of how these new concepts and technologies would change the existing air traffic control (ATC) system. Furthermore, there was no single-source document available to help us leverage our awareness and understanding of the sweeping changes that some are calling the most significant "paradigm shift" in aviation history.

In this book, Vince Galotti has given us a comprehensive, easy to understand, road-map that helps bridge the gap in our ability to visualize the impact of moving from the 60 year old terrain-based ATC system to the CNS/ATM environment of the 21st century. Unlike much of the information available on the new CNS technologies, this book was not written by an engineer for engineers. It is clear and concise and moves the reader from topic to topic with efficiency and without trying to "sell" any particular argument along the way.

The United States' National Airspace System (NAS) is one of the world's most complex systems. It engages thousands of people, and billions of dollars are invested in aircraft, facilities and equipment. The importance of the CNS/ATM system will be measured by its ability to move people safely and efficiently all over the world. The fact that the NAS and ATC systems throughout the world will be vastly changed in the 21st century creates an urgent need for aviation professionals everywhere to gain a profound knowledge of how all the CNS/ATM components, systems and people must be integrated to provide the safety and cost-benefits that are essential for the continued growth of aviation transportation.

From the perspective of an aviation professional who has worked within the ATC system and then moved into the international arena, Vince Galotti writes with the authenticity of someone who not only knows his subject, he has "been there, done that" and can therefore, explain, compare and give dimension to his topics. If you have any investment in the future of aviation, I highly recommend you read this book. When you have finished reading it, you may find a quotation attributed to Sir Winston Churchill appropriate: "It is not the end. It is not the beginning of the end. However, it is the end of the beginning".

Marvin L. Smith Ed.D

Preface

Excerpt from President Franklin Roosevelt's message to the Opening Plenary Session of the Chicago Conference in 1944

> As we begin to write a new chapter in the fundamental law of the air, let us all remember that we are engaged in a great attempt to build enduring institutions of peace. These peace settlements cannot be endangered by petty considerations or weakened by groundless fears. Rather, with full recognition of the sovereignty and judicial equality of all nations, let us work together so that the air may be used by humanity, to serve humanity (Department of State, United States Government, 1944).

I would like to thank God; all those who helped me through the difficult times in my life, some of which took place during the writing of this book, those who continue to help me and those who will help me later, you know who you are. I would like to thank my son who has brought me the happiest moments in my life and taught me the meaning of the word Love, and my parents for their constant love, support and encouragement. I would also like to acknowledge those who encouraged me to take on this task and saw to it that the necessary permission was granted: Chris Eigl, the ICAO Representative of the European and North Atlantic Office, who supported me, proofread and offered encouragement; Manfred Krull, the Deputy Director of the ICAO Technical Cooperation Bureau who also proofread, supported and encouraged me; Dr. Philippe Rochat, the Secretary General of ICAO for approving this project without hesitation; Mr. Norm Solat and others for proofreading and all of the talented people inside and outside of ICAO who had the vision and the ability to create the FANS CNS/ATM systems concept and then develop the technical material that I would later use in creating this text. They continue to work to further the goals of international civil aviation and I wish them well and am honoured to be a part of that community. I would like to thank the professors at Embry-Riddle Aeronautical University who taught me so much about research, writing, referencing, etc., during my time there, especially those on my Thesis Committee, and particularly, Dr. Marvin L. Smith of the Aeronautical Science Department, who also acted as mentor during my university years and continues to prop-up my ego. Finally, I would like to thank Natasha Eguer in Paris for her friendship, support and encouragement and for helping with the typing after working hours, Helen Manentis in Montreal for her support and help and Lynn McGuigan for supporting me with her extraordinary word processing skills and graphic design abilities, but most of all for her determination and enthusiasm toward this project. Finally, I would like to thank Mr. Philippe Domogala for contributing Chapters 9 and 10, which are based on his exceptional personal experience and knowledge.

PART A

HISTORICAL PERSPECTIVE

The Boeing 777 flight deck with new display applications, including those necessary to manage ATM applications. (Picture provided courtesy of Boeing Commercial Airplane Group.)

1. Introduction to the Future Air Navigation System (FANS)

Introduction

On September 5th, 1991, 450 representatives from 85 nations and 13 international organizations gathered at the headquarters of the International Civil Aviation Organization (ICAO) in Montreal, Canada to consider and endorse a concept for a future air navigation system (FANS) that would meet the needs of the civil aviation community over the next century. The concept, known as FANS, eventually came to be known as Communications, Navigation, Surveillance/Air Traffic Management (CNS/ATM) systems. CNS/ATM involves a complex and interrelated set of technologies, dependent mainly on satellites.

In addition to the technical elements, there are also institutional, economic and human factors aspects that must be considered as nations prepare for transition to the new systems. Many of the technical issues and other aspects of CNS/ATM systems are discussed in this book. This introductory chapter provides an overview of the historic events leading up to and through the development of CNS/ATM systems, and also generally describes the CNS/ATM system elements and expected benefits.

The Special Committee on Future Air Navigation Systems - (FANS)

The future air navigation systems concept, which has been developed by ICAO and has now evolved to become known as the ICAO Communications, Navigation, Surveillance and Air Traffic

> Management systems, is essentially the application of today's high technologies in satellites and computers, data links and advanced flight deck avionics, to cope with tomorrow's operational needs (Kotaite, 1993, p. 5).

Having considered the steady growth of international civil aviation preceding 1983, taking into account statistical forecasts at the time, and perceiving that new technologies were on the horizon, the Council of ICAO considered the future requirements of the civil aviation community (see Figure 1.1). As an outcome of their extended deliberations, the Council determined that a thorough analysis and reassessment of the ideas, technologies and institutional arrangements that had so successfully served civil aviation for over fifty years, was in order. In further predicting that the systems and procedures supporting civil aviation, because of their inherent limitations, were incapable of dealing with expected growth, the Council took an important decision at a pivotal juncture; a decision that is already coming to be seen as significant and timely, if not historic. The decision of the Council was to establish the Special Committee on Future Air Navigation Systems (FANS). Among other things, the FANS Committee was tasked with studying, identifying and assessing new technologies, including satellite technology, and making recommendations for the future development of air navigation for civil aviation over a period of the order of twenty five years (see Appendix C for a description of the work programme of the FANS Phase I Committee).

FANS membership was initially drawn from twenty two ICAO Contracting States and several international organizations, eventually being comprised of members from some forty nations and international organizations in what would become a collaborative effort on a global scale (O'Keefe, 1993).

Four full committee meetings were held over a period of four years, supported by valuable preparatory work performed by task forces, working groups and related institutions and organizations. The FANS Committee completed its work in October of 1988 and presented a comprehensive final report to the President of the ICAO Council (ICAO, 1988a).

Proving the foresight of the Council to be well placed, the FANS Committee determined that the shortcomings inherent in the communication, navigation and surveillance systems of the time, were indeed incapable of supporting the future needs of civil aviation. Although the effects of these shortcomings were not the same for every part of the world, the FANS

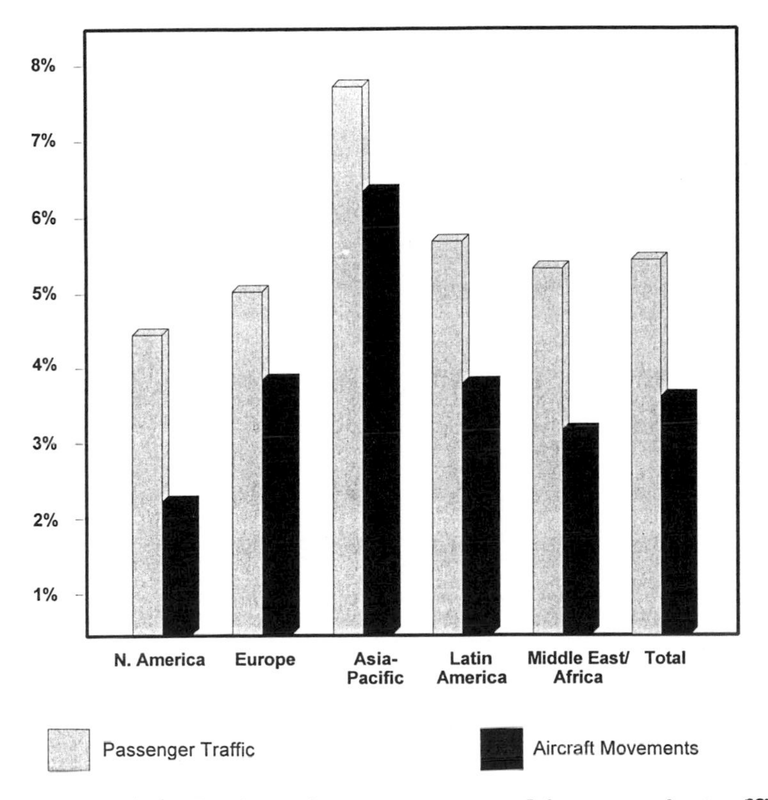

Figure 1.1 Projected average annual increase in traffic and movement (1992-2010)

Committee had established that, in some form, these shortcomings would inhibit the further development of air navigation almost everywhere.

The FANS Committee determined that it would be necessary to develop new systems that would overcome limitations and allow air traffic management (ATM) to develop on a global scale. The future system would be expected to evolve and become more responsive to the needs of those operating in the system, and whose economic health is directly related to the efficiency of that system.

Winding down its efforts in 1988, the FANS Committee concluded that satellite technology offered the only viable solution to overcome the shortcomings of the present systems and to meet the future needs of the international civil aviation community. Indeed, the fourth and final report of the FANS Committee made clear that satellite based systems would be the key to worldwide improvements in civil aviation (ICAO, 1988a).

A further encouraging aspect of the Committee's work was the finding that the economic impact resulting from the wide scale use of such technology could result in annual benefits to the world aviation community of between 5.2 and 6.6 billion U.S. dollars (ICAO, 1988b).

The Special Committee for the Monitoring and Coordination of Development and Transition Planning for the Future Air Navigation System (FANS Phase II)

> This future air navigation system, which departs significantly from the conventional one, requires global implementation (Kotaite, 1992, p. 14).

Having completed the task assigned to it by the Council, which was the development of a concept or blueprint for future CNS systems, the FANS Committee recognized that its concept would require an extraordinary change in the way that sovereign nations have historically developed and implemented their civil aviation systems, even if in an evolutionary and incremental manner. This was because future systems would be based mainly on satellites, from which coverage would extend over the airspace of several nations. And unlike their terrestrial counterparts, satellite CNS systems would provide near global coverage of the earth.

It was recognized even in these early stages of conceptual development that the special characteristics of CNS/ATM technology would present the international civil aviation community with some novel considerations. Traditionally, individual nations procured, certified, operated and maintained their own ground based portions of ICAO defined air navigation systems, which were then used in their own territories for national purposes, but to a greater extent, for international civil aviation purposes in accordance with a wider ranging international air navigation plan established by ICAO. This would no longer be the case with satellite based systems which, by necessity, and by their very nature, would be established as a shared resource (Featherstone, 1993). Furthermore, civil aviation administrations of individual nations have historically formed their air traffic systems to suit their own needs in terms of individual, military and political interests. As a result, there are serious and ever present interface problems which inhibit the smooth flow of information and air traffic. As the new systems will be global by design, each individual system will have to be integrated with the global system. Thus, all nations would have to adopt a global view (O'Keefe, 1991) if the new systems are to be successfully implemented.

The FANS Committee recognized that the evolution of ATM on a global scale using CNS concepts would require a multi-disciplinary approach because of the close inter-relationship of its many elements (ICAO, 1988a). Understanding that coordination and the institutional issues could eventually arise with their concept for the future, and realizing that planning would have to be carried out at the worldwide level, the FANS Committee recommended to the ICAO Council in its final Report:

> that a new committee be established urgently to advise on the overall monitoring, coordination of development and transition planning to ensure that implementation of the future CNS system takes place on a global basis in a cost-effective manner and in a balanced way between air navigation systems and geographical areas (1988a, p. 5-6).

In July 1989, the ICAO Council, acting on the recommendation of the FANS Committee, established the FANS Phase II Committee (see Appendix C for a description of the work programme of the FANS Phase II Committee).

The FANS Phase II Committee determined that the goals of the future system should include:

> enhanced safety, accommodation of the full range of aircraft types and airborne capabilities, improvement of provision of information to users (weather, traffic situation, availability of facilities), flexible airspace management, efficient use of airspace, increased user involvement in ATM decision making (through air-ground computer dialogue), and creation, to the extent possible, of a single continuum of airspace where boundaries are transparent to the users (ICAO, 1993a, p. 3).

In October of 1993, after ten combined years of effort, begun by the first FANS Committee, the FANS Phase II Committee completed its work. Both phases of the FANS Committee recognized that implementation of related technologies and expected benefits would not arrive overnight, but would rather evolve over a period of time, depending upon the region, the present aviation infrastructures of individual nations and the overall requirements and needs of the aviation community.

The FANS Phase II Committee also agreed that much of the technology they were considering, was already becoming available and recognized that transition to CNS/ATM systems should begin by gathering information on prior experience, and accruing early benefits using available technologies.

It is widely accepted today that the work of these two committees will determine the shape of international civil aviation well into the next century.

The Tenth Air Navigation Conference

On September 5th, 1991, 450 representatives from 85 ICAO Contracting States and 13 international organizations gathered at ICAO Headquarters in Montreal, Canada to consider and review, in some detail, the work achieved by the FANS Committees. At the same time, these high level civil aviation representatives of the nations of the world considered and made recommendations, based on the work of the FANS Committees, as well as on their own discussions, which would assist the worldwide civil aviation community towards shaping its future.

These representatives met for just over two weeks. The results of their work was the final report of the Conference which encapsulates a set of universally agreed recommendations covering the full spectrum of CNS/ATM activities, that will offer guidance and direction to the world civil aviation community as they begin planning and implementation of the technical aspects of CNS/ATM systems (several of these recommendations are reproduced at Appendix D). Among other things, the Tenth Air Navigation Conference further considered:

* The individual elements of CNS/ATM systems;
* The capabilities of this concept for overcoming shortcomings;
* Institutional aspects;
* Air traffic management;
* Global transition planning;
* Safety, technical, operational and economic aspects including cost-effectiveness and penalties associated with non-implementation;
* Transition planning; and,
* Coordination of implementation.

Perhaps, however, the single most important item on the agenda of the Tenth Air Navigation Conference was the endorsement of CNS/ATM systems. Quite predictably, the conference reached a consensus on this need and also agreed upon the urgency in progressing activities associated with implementation of the global system. In fact, it was felt that planning for the future required a clear confirmation by the international civil aviation community that CNS/ATM systems were the accepted worldwide approach for the future. The conference was quite logically taking on an historic quality by the time the delegates began reaching their final conclusions. In recognizing the timeliness of their endorsement, it was noted that:

> the exploitation of technology, especially in space and ground segments, had progressed to the point where airspace users were beginning to derive significant benefits. Maritime transport, for example, had already made extensive use of satellites for navigation and communication.... The pace of change was quickening, and aircraft operators and air traffic services (ATS) providers were committing themselves more and more to the available technology (ICAO, 1991a, p. 9-2)

In concluding that their endorsement was necessary, these representatives of the nations of the world recognized that CNS/ATM systems offered solutions to the identified shortcomings, while also taking full advantage of existing and foreseen technologies.

A brief look at CNS/ATM

The four main elements of CNS/ATM systems are summarized below, and are dealt with in detail in later chapters of this text:

Communications

In future CNS/ATM systems, the transmission of voice will continue to take place over existing very high frequency (VHF) channels; however, these same VHF channels will increasingly be used to transmit *digital data* (data link is discussed in detail in Chapter 4). Satellite data and voice communications, capable of global coverage, will also be introduced. The regular use of data transmission for ATM purposes will introduce many changes in the way that communication between air and ground takes place, and at the same time offer many possibilities and opportunities.

The Secondary Surveillance Radar (SSR) Mode S (Mode S radar is discussed in detail in Chapter 5), which will be used for surveillance in high density airspace, will also have the capability of transmitting digital data between air and ground, and an aeronautical telecommunications network (ATN) will provide for the interchange of digital data between end users over dissimilar air-ground and ground-ground communication links (the ATN is discussed in detail in Chapter 3).

The benefits expected from the future communications systems lie in the fact that they will allow more direct and efficient linkages between ground and airborne *automated* systems while offering the possibility of pilot/controller communication taking place via data link, using data displays instead of voice. In fact, digital data link can be seen as the key to development of new ATM concepts leading to the achievement of real benefits.

Navigation

Improvements in navigation will include the progressive introduction of area navigation (RNAV) capabilities along with global navigation satellite systems (GNSS). These systems will provide for worldwide navigational coverage and will be used not only for worldwide enroute navigation, but also, initially, for non-precision approaches and eventually for precision approaches. GNSS will provide a high integrity, high accuracy, worldwide navigation service. The implementation of these systems will enable aircraft to navigate in all types of airspace, in any part of the world, using simple on-board avionics to receive and interpret satellite based signals, offering the possibility for many nations to dismantle at least a portion of their aviation ground infrastructures (GNSS is discussed in detail in Chapter 4).

Surveillance

It is expected that traditional SSR modes will continue to be used in terminal areas, along with the gradual introduction of Mode S in both terminal areas and high density continental airspace. The major breakthrough however, will be with the implementation of automatic dependent surveillance (ADS), which will allow aircraft to automatically transmit their position, and other data, such as heading, speed and other useful information contained in the flight management system (FMS), via satellite or other communication link, to an air traffic control (ATC) unit where the position of the aircraft could be displayed somewhat like it would be on a radar display (ADS is discussed in detail in Chapter 5). ADS can also be seen as a communications application that represents the true merging of communications and navigation technologies, and, along with ground system automation enhancements, will allow for the introduction of enormous benefits for ATM, especially in oceanic airspace where aeronautical mobile satellite services (AMSS) will also be made use of. Software is currently being developed that would allow this digital data to be used directly by ground computers to detect and resolve conflicts. Eventually, this could lead to clearances being negotiated between airborne and ground based computers with little or no human intervention.

Benefits would be derived quickly through ADS in oceanic and some continental areas that currently have no radar coverage.

Air Traffic Management (ATM)

If we combine the areas of communications, navigation and surveillance and all of the expected improvements, the overall main beneficiary is likely to be ATM. More appropriately, the advancements in CNS technologies will all serve to support ATM. When referring to ATM in the future concept, much more than just air traffic control is meant. In fact, ATM refers to a system's concept of management on a much broader scale, which includes ATS, air traffic flow management (ATFM), airspace management (ASM) and flight operations (ATM is discussed in detail in Chapter 6).

The combined benefits expected from the new and available CNS technologies, together with the increasing use of automation, will lead to an improved ATM system. Figure 1.2 shows how the expected benefits of CNS technology will integrate to benefit ATM.

Conclusion on CNS/ATM systems

CNS/ATM systems will improve the handling and transfer of information; extend surveillance using ADS; and navigational accuracy would be improved through GNSS. This will lead to reductions in separation between aircraft, allowing for an increase in airspace capacity.

Advanced CNS/ATM systems will also see the implementation of ground based intelligent type processors to support increases in traffic. These ground based systems will exchange data directly with FMSs aboard aircraft, through data linking, making use of the ATN. This will benefit the ATS provider and airspace user by: enabling improved conflict detection and resolution through intelligent processing, providing for the automatic generation and transmission of conflict free clearances, as well as offering the means to adapt quickly to changing traffic requirements. As a result, the ATS provider will be better able to accommodate an aircraft's preferred flight profile and help aircraft operators to achieve reduced flight operating costs and delays.

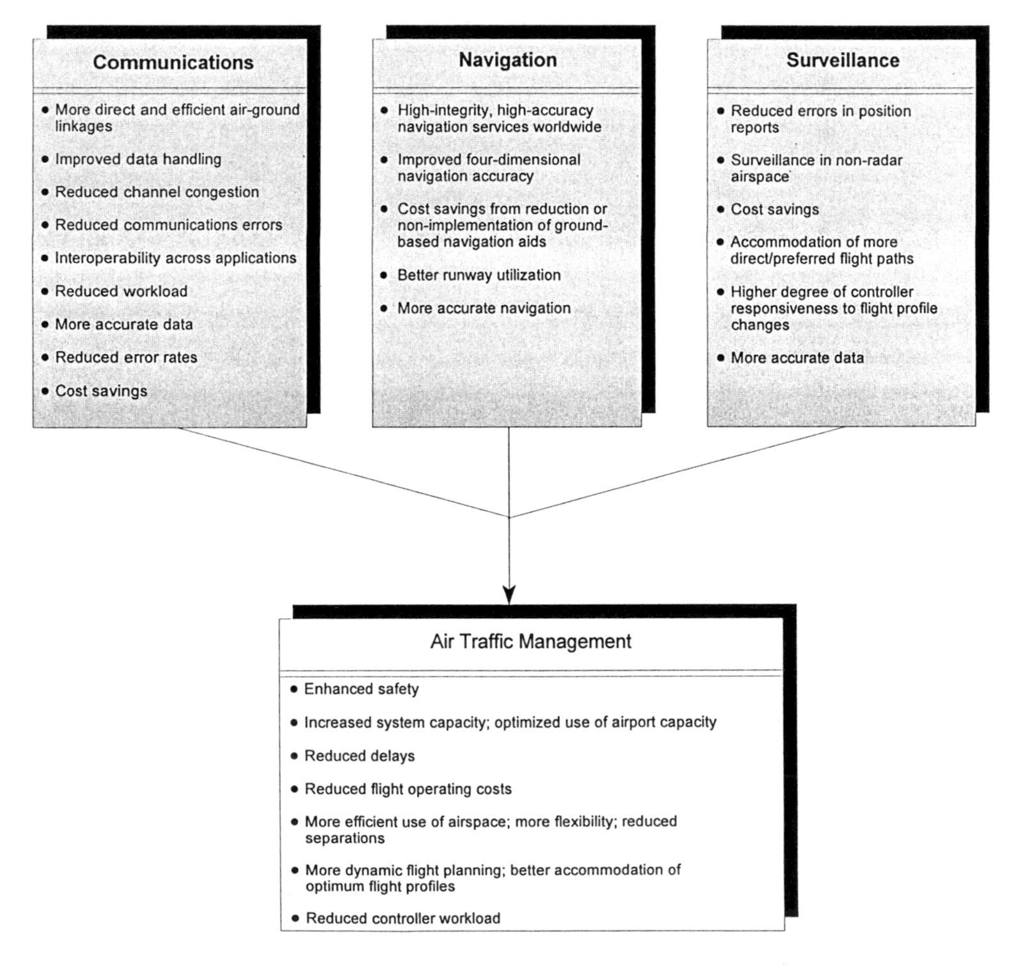

Figure 1.2 Overall ICAO CNS/ATM systems benefits: A high level view

A glance at the future

According to the International Coordinating Council of Aerospace Industries Association (ICCAIA, 1991), "over 85 per cent of today's wide-bodied aircraft are delivered with satellite communications provisions installed, and a new aircraft equipped to make use of CNS/ATM technology enters service every other working day". That statement was made in 1991. Today, the trend continues to be towards ever increasing numbers of aircraft being equipped to make use of satellite technology.

It becomes clear that the primary benefits to be had from implementation of CNS/ATM systems will come only after both providers and users of the air navigation system commit themselves to implementation of the major component elements (see Figure 1.3). For example, the core of the benefits of the future system will be derived from automation intended to reduce or eliminate constraints imposed on ATM operations (Andresen, 1991). Most of the automation however, is dependent on developments in computer software. It is certain that advances in software can only be fully exploited if the other main elements of CNS/ATM technology are in place. As an example, work is rapidly progressing on computer programmes that will assist the air traffic controller in identifying and resolving air traffic conflicts, or otherwise assist controllers in performing their duties. To take full advantage of such programmes, ground computers will need access to navigational and other data from computer systems on board aircraft, such

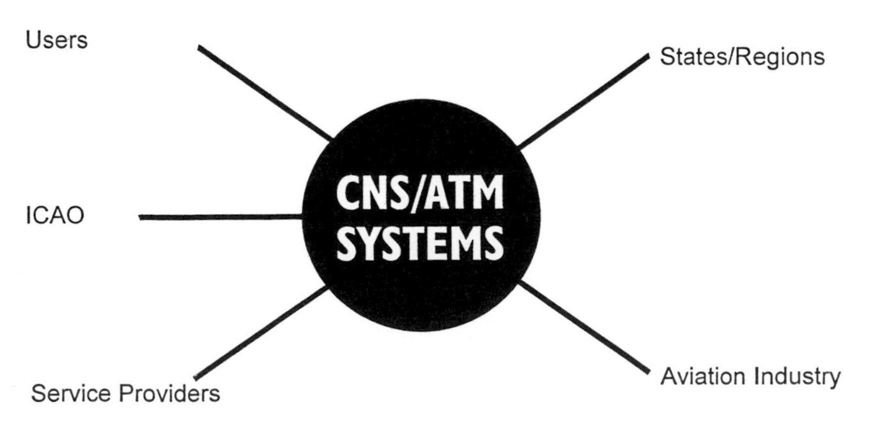

Figure 1.3 Entities involved in transition to CNS/ATM systems

as from the FMS, thereby forming a more close-knit relationship between ground and air. For the rapid and reliable transmission of this data to take place between ground and airborne computers however, data linking capabilities for CNS/ATM systems, will be necessary. Communication via data link and the use of automated processing are therefore among the key elements of CNS/ATM systems. It is no wonder that airlines strongly encourage the use of data link and other CNS/ATM related technology.

Benefits for the airlines

> The FANS 1 upgrade for the Boeing 747-400 flight management system will ultimately be the key to the world's first operational integration of communication, navigation and surveillance (Aleshire, 1994a).

A significant proportion of today's air carrier aircraft operate quite sophisticated aircraft with on-board capabilities that generally far exceed what is now used by the ATC system (International Air Transport Association [IATA], 1991). For this reason the airlines strongly endorse implementation of CNS/ATM technology. CNS/ATM technology will allow substantial savings through reduced operating costs for airlines because of increased opportunities to operate in conformance with the most efficient flight profiles as well as because of an expected reduction in delays as the airspace is used more efficiently.

In describing the recognizable advantages to early implementation of CNS/ATM technology and installation on United's Boeing 747-400 aircraft, Capt. W. Aleshire (1994a) of United Airlines expects: ATC separation standards over the Pacific Ocean will be reduced; early access to the vast airspace over the Russian Federation; introduction of 1,000 ft. vertical separation above 29,000 ft. (the current standard is 2,000 ft.) and the opening of new routes, such as Chicago - Hong Kong; an overall enhancement of safety along with a concurrent reduction in communication and other errors. In addition to the benefits, interim introduction of various elements of CNS/ATM systems will not require major investments on the part of the airlines.

The first FANS equipped Boeing 747-400 was certificated in June 1995, and entered into service on Pacific routes the following month. Not far behind Boeing, Airbus and McDonnell Douglas are working on similar packages for their respective long-haul aircraft, the A 330/340 and the MD-11 (Norris, 1994a).

Benefits for the nations that provide the air navigation infrastructure

For the nations of the world, responsible for providing and maintaining the extensive ground infrastructures necessary to support civil aviation, which include the provision of ATC and other supporting air traffic services, a reduction in the cost of operating and maintaining those infrastructures is expected, as the traditional ground systems become obsolete and satellite technology is increasingly employed. They will also benefit from enhanced safety.

CNS/ATM technology provides a timely opportunity for developing nations to enhance their systems. Many of these nations have large areas of available, but unusable airspace, mainly because of the expense of purchasing, operating and maintaining the necessary ground infrastructures. CNS/ATM systems will afford them opportunities to modernize inexpensively. This will provide them the means to accept and handle large, sophisticated aircraft, so vital to the economic health of those nations.

Much of the CNS/ATM technology is already available, can use existing ground networks and will offer lower capital and maintenance costs than current systems.

General aviation

General aviation and utility aircraft will find increasing access to avionics equipment that will allow them to operate in flight conditions as well as into and out of airports that they would normally have been prohibited from using because of the cost of equipping their aircraft.

Furthermore, many remote areas that are currently inaccessible to most general aviation aircraft because of their inability to communicate or safely navigate over them, would become more accessible thanks to satellite communications, navigation and surveillance.

Conclusion (glance at the future)

To be sure, there are still quite a number of issues that will have to be dealt with and overcome if all of the possible benefits of CNS/ATM systems are to be fully exploited. This will involve the combined effort and goodwill of both ATS providers and those that operate within the system and who will take advantage of these technologies. This may not be easy. For instance, there are major institutional questions that will eventually have to be addressed, which include funding, ownership of facilities, access to satellite equipment and technical issues related to availability, reliability and integrity, many of which are being addressed by the international civil aviation community through ICAO. Human factors issues concerned with the man-machine interface and related labour issues will also have to be thoroughly examined. Overall, however, CNS/ATM systems are being viewed upon favourably by those who have authority over and who operate within the air navigation system, and the reality of early implementation of various CNS/ATM system elements is becoming widely accepted by the aviation community. Figure 1.4 shows the differences between current CNS systems and the future system.

> The aerospace industry wishes to work cooperatively to promote the FANS concept: a safe and reliable, efficient and economical worldwide global air transportation management system that really can make a meaningful difference in satisfying capacity demands (ICCAIA, 1991, p. 15).

Recapping the major points

The international civil aviation community is continuing its consistent evolutionary process of adjustment, expansion and technological change. In 1983 a critical juncture was reached and the aviation community embarked on a course that began with the recognition by the Council of ICAO that a thorough analysis and reassessment of the ideas, technologies and institutional arrangements that had so successfully served civil aviation for fifty years, was necessary. A series of events followed:

* The establishment by the ICAO Council of the Special Committee on Future Air Navigation Systems, to address the critical issues;

* The development of CNS/ATM systems that would be based mainly on satellite technology;

* The creation of a successor committee in 1988 - the Special Committee for the Monitoring and Coordination of Development

Figure 1.4 CNS system evolution

and Transition Planning for the Future Air Navigation System (FANS Phase II) - to deal with implementation and transition issues associated with CNS/ATM systems;

* The acceptance of the work of both FANS Committees by the world community in 1991 at the ICAO Tenth Air Navigation Conference and the formal endorsement of CNS/ATM systems for global implementation.

CNS/ATM systems will improve the handling and transfer of information mainly by making use of data linking techniques; it will extend surveillance of aircraft using ADS; and navigational accuracy will be improved through global navigation satellite systems.

Today, many nations, airlines and aircraft manufacturers are beginning with early implementation of CNS/ATM technologies.

Questions and exercises to expand your knowledge

1) This chapter mentions inherent limitations in current aviation systems. Could these limitations have something to do with the current system elements not having the capability of adapting to other technological advancements. For instance, present communications systems evolved to transmit voice messages and not data messages. Why would it be advantageous to be able to transmit data also, as opposed to voice only? Why would it be important to retain some of the voice capability rather than evolve toward a data only environment?

2) Over the past fifty years, many developments have taken place which include the turbine engine, the turbofan engine, growth in the size, speed, efficiency and environmental acceptability of aircraft and advancements in computer technology. The older technologies, which served the world community so well for so long, are in danger of losing their effectiveness and capability in coping with the realities of today and tomorrow. Some of the reasons for this are listed below. Expand upon these:

   * Economic realities of the times;
   * The cost of doing business;
   * Efficiency requirements;
   * Free trade and competition;
   * Opportunity;
   * Implementation.

3) Although covered in more detail in later chapters, explain how you would expect advancements in satellite technology to affect the way that we:

   * Communicate;
   * Navigate;
   * Monitor the progress of flights.

4) How do you think advancements in computer technology (hardware and software) or, more generally, automation, could impact the elements of the aviation infrastructure listed in question three?

5) Explain the special characteristics of satellites that would require extensive cooperation and coordination between nations?

An oceanic air traffic management flight data display system. A window also provides automatic dependent surveillance (ADS) aircraft position data. (Picture provided courtesy of CAE Electronics Ltd.)

2. The International Civil Aviation Organization (ICAO)

Introduction

> Commercial aviation is the safest form of transportation in the world because of the foresight and perseverance of the delegates from 54 nations that attended the civil aviation conference in Chicago from November 1 to December 7, 1944.... The conference delegates drafted a framework of safety and technical standards that remains the foundation of civil aviation and air navigation, and set in motion the creation of the International Civil Aviation Organization (Jackman, 1994, p. 49).

Have you ever wondered how an aircraft manufactured in the United States or Russia can fly from Saudi Arabia to London or New York or from Kiev to Tokyo, or an aircraft manufactured in Brazil can fly from Sweden to France with a German or Norwegian pilot at the controls? How about common standards and accepted practices and regulations for licensing and qualification procedures of the pilots; certification of airframes, engines, communications and avionics equipment? Are they the same in every nation of the world? What responsibility do individual nations have to install navigational facilities and to provide services and other aviation infrastructure? What kind of procedures do air traffic controllers follow in different countries and how would a pilot know what these are? In what language do the pilots and controllers speak to each other? What about customs, security, the carriage of dangerous goods, aircraft registration and markings, noise and environmental regulations? Are there universal reaction times and procedures for fire fighters and rescue operations? Or universal standards for airfield and approach lighting and markings? These are just a few of the many elements that must be looked at when considering the complexities of international civil aviation.

To understand the answers to these questions, one would have to understand the workings and functions of ICAO. And to fully appreciate and understand CNS/ATM systems as briefly described in the previous chapter, and how the nations of the world could come together and agree upon a common set of ideas and a vision for the future air navigation system, one would have to have a grasp of the underlying realities and historical events associated with international harmonization and globalization, and of the adoption and agreement of international standards. Furthermore, developing the concept for CNS/ATM marked only the beginning of an enormous task. Next, the international community will have to transition to and implement the new systems in a cooperative and coordinated way.

This chapter provides a brief history of internationalism in civil aviation so that you will gain an awareness of how CNS/ATM systems came into being. It also describes the workings and functioning of ICAO and how these are related to CNS/ATM. The very fact that international civil aviation functions so well and the concept of CNS/ATM could be developed, endorsed and eventually implemented by the international civil aviation community at the worldwide level, is a testimony to the success of internationalism in aviation. Figure 2.1 gives an example of ICAO published international signs to provide guidance to persons at airports.

A historical synopsis of internationalism in aviation

Freer (1986a) traces the roots of internationalism as it came to be embodied in ICAO, back to 5 June 1783, in Annonay, France, when Joseph and Etienne Montgolfier launched their hot air balloon, the Global Aerostatique. Over the next few years, the need for some kind of regulation became necessary for the governments and police authorities of France, and several other European Nations. These were usually simple measures of a local nature. These vehicles grew in size and sophistication, covering even greater distances, eventually leading to the first truly international flight in 1785 when the first manned hot air balloon crossed the English Channel.

Figure 2.1 International signs to provide guidance to persons at airports

Over the next one hundred years, man persisted with his exploration in pursuit of more advanced and proficient flying machines, experimenting with hot air, propellers, propulsion devices and dirigibles. Along with this success, governments became increasingly aware of the need for bilateral and international understandings.

The birth of international civil aviation

The seventeenth of December 1903 was a historic date in aviation as the Wright brothers succeeded in launching the first powered, heavier than air flight at Kitty Hawk, North Carolina. According to Freer (1986a), the Wright brothers flight also marked the birth of international civil aviation as we know it today.

In the succeeding years there was an explosion of heavier than air experimentation and activity. The aeroplane eventually developed into a major instrument of transportation bringing with it international problems; problems of coordination of techniques and laws, of technical and economic issues, all of which were well beyond any of the individual governments of the time to solve.

In the intermittent years before 1910, several international gatherings took place which attempted to deal with regulatory issues mainly having to do with aerial crossings of international frontiers.

In 1910, only seven years after the first heavier than air flight, the French Government invited twenty one European Nations, of which eighteen attended, to meet in a diplomatic conference held in Paris in order to address subjects related to the regulation of air navigation in Europe. The conference attempted to draft an international air navigation convention. And although the delegates could not agree on a definitive text for a convention, the conference did come to be seen as the first important and truly significant international gathering of its kind. A number of important principles governing aviation, laying the framework for future agreements, were agreed to in Paris in 1910.

World War I and the Paris Convention of 1919

The technical developments emanating out of World War I created a new situation as hostilities ended, especially regarding the safe and rapid transport of people and goods over long distances. It was therefore no surprise that immediately after the end of the war, civil aviation resumed the rapid growth that it had experienced over the first fourteen years of the century (Berger, 1984).

The year 1919 was a significant one for civil air transport with the establishment of the first commercial air service between London and Paris. Two crossings of the Atlantic Ocean were also made in that year and several other historical events were recorded. The most significant event regarding international civil aviation however, was the creation of the Aeronautical Commission of the Paris Peace Conference (ICAO, 1947) which would eventually lead to the International Air Convention. In its 43 articles, the Convention dealt with such issues as the general principles for regulating air navigation, which included nationality of aircraft, admission of aircraft above the territories of Contracting States and general arrangements to be made by all Contracting States in order to develop aviation and the handling of disagreements and other technical, operational and organizational aspects of civil aviation. Also in the convention was the provision for an International Commission for Air Navigation (ICAN). Over the next few years, ICAN established itself as a focal point for government and industry coordination and as a recognized aviation authority (Freer, 1986b). The Convention also took over all of the principles which had already been formulated by the conference held in Paris in 1910.

Although ICAN had accomplished a useful purpose, it was never truly able to deal with international aviation on a worldwide basis. The United States, the Soviet Union, Germany, China and most of Latin America did not participate. The reasons for lack of participation ranged from the forces of political isolationism in the United States to a perceived negative impact of a few of the articles of the convention, to a general lack of interest on the part of the Soviet Union (Department of State, United States, 1948).

A group of nations in the Western Hemisphere, including the United States, met in Lima in 1937 in an attempt to establish a second international civil aviation body under the Pan American Convention for Air Navigation drawn up in Havana, Cuba in 1928. This permanent American aeronautical commission, however, was never formally constituted, neither was it considered suitable to be used on a worldwide basis.

World War II and the Chicago Conference

With the beginning of World War II in 1939, the carriage of passengers and freight by air, although growing rapidly, was still a relatively minor means of international transport. Military operations over the course of the war, however, promoted the aeroplane to a major component of the world's transportation system. The Royal Air Transport Command for example, which took over a small civilian flying organization ferrying aircraft from North American factories to the United Kingdom, expanded into a worldwide network which carried passengers and freight to all military theatres. Similarly, the United States Army Air Transport Command and the Navy Air Transport Service became major elements of the overall United States' military power (ICAO, 1951). A network of air routes and military airfields were established around the world. Thus, the pressure of the military of a world at war compressed a quarter century of normal peacetime aviation development into a few years.

Wartime cooperation overcame many political barriers, but there were many obstacles, political and technical, that still had to be overcome if the fruits of the wartime enterprise were to benefit international civil aviation. The following main issues were the most prevalent among many:

* The problem of commercial rights;

* What arrangements had to be made for the airlines of one country to fly into and through the territories of another;

* What action should be taken to minimize the legal and economic conflicts that might come with peacetime flying across national borders; and,

* What could be done to maintain war-created air navigation facilities.

With the end of World War II in sight, and with the above issues outstanding, the Government of the United States conducted exploratory discussions with other allied nations during the early months of 1944. On the basis of these talks, invitations were sent to fifty five allied and neutral nations to meet in Chicago in November 1944 (ICAO, 1953). The following is an excerpt of the invitation letter: (the full text of the invitation to the Chicago Conference is at Appendix B).

> The approaching defeat of Germany, and the consequent liberation of great parts of Europe and Africa from military interruption of traffic, sets up an urgent need for establishing an international civil air service pattern on a provisional basis at least, so that all important trade and population areas of the world may obtain the benefits of air transportation as soon as possible, and so that the restorative processes of prompt communication may be available to assist in returning great areas to processes of peace (Department of State, United States Government, 1944).

The delegates of these nations met for five weeks to make arrangements for the establishment of world air routes and services, and to set up an interim council to collect, record and study data concerning international aviation in order to make recommendations for its improvement. The Conference would also be called upon to discuss the principles and methods to be followed in the adoption of a new aviation convention. This convention established an organization designed to foster and to guide international civil aviation. Its purpose is set forth in the Preamble to the Convention:

> *Whereas* the future development of international civil aviation can greatly help to create and preserve friendship and understanding among the nations and peoples of the world, yet its abuse can become a threat to the general security; and,
>
> *Whereas* it is desirable to avoid friction and to promote that cooperation between nations and peoples upon which the peace of the world depends;
>
> *Therefore* the undersigned governments having agreed on certain principles and arrangements in order that international civil aviation may be developed in a safe and orderly manner and that international air transport services may be established on the basis of equality of opportunity and operated soundly and economically;
>
> Have accordingly concluded this convention to that end (ICAO,1953, p. 8).

In a message to the Opening Plenary Session of the Chicago Conference, President Franklin Roosevelt clearly set forth the fundamental aims of the international community, reflecting their hopes for peace and renewal. The following is an excerpt from his message:

> As we begin to write a new chapter in the fundamental law of the air, let us all remember that we are engaged in a great attempt to build enduring institutions of peace. These peace settlements cannot be endangered by petty considerations or weakened by groundless fears. Rather, with full recognition of the sovereignty and judicial equality of all nations, let us work together so that the air may be used by humanity, to serve humanity (Department of State, United States Government, 1944).

The permanent body charged with the administration of the principles as laid out by the Chicago Conference is the International Civil Aviation Organization (ICAO). The Convention provided that ICAO would not come into being until the Convention was ratified by twenty two nations.

Realizing that ratification of an international agreement would require sufficient time for the various governments to review and enact the necessary legislation, the delegates provided for a provisional ICAO (PICAO) to operate until the permanent body was created. Over its twenty months of life, PICAO was able to take action to provide and maintain the facilities and services necessary for the operation of air services across national borders. Much of their work involved the drafting of recommendations for standards, practices and procedures designed to ensure the safety, regularity and efficiency of international air transport (ICAO, 1958).

The aims and objectives of ICAO as laid out by the Chicago Conference are as follows:

* Ensure the safe and orderly growth of international civil aviation throughout the world;

* Encourage the arts of aircraft design and operation for peaceful purposes;

* Encourage the development of airways, airports and air navigation facilities for international civil aviation;

* Meet the needs of the peoples of the world for safe, regular, efficient and economical air transport;

* Prevent economic waste caused by unreasonable competition;

* Insure that the rights of Contracting States are fully respected and that every Contracting State has a fair opportunity to operate international airlines;

* Avoid discrimination between contracting States;

* Promote safety of flight in international air navigation;

* Promote generally the development of all aspects of international civil aeronautics.

In 1947 an agreement was signed by the President of the ICAO Council and the United Nations Secretary General by which ICAO became a specialized agency in relationship with the United Nations, and thereby became a member of the United Nations family.

The ICAO Convention

One of the lasting achievements and most successful undertakings of ICAO, has been its work towards uniformity of standards and procedures, which in itself is one of the principal reasons for the success of international civil aviation over the past fifty years. In fact, each signatory State "undertakes to collaborate to secure the highest practicable degree in uniformity of regulations and standards, procedures and organization relating to aircraft, personnel, airways...in which such uniformity would be likely to facilitate and improve air navigation" (ICAO, 1951).

The largest area of controversy in the Convention had to do with the question of controlling the right of one nation's airlines to carry traffic between the airports of two other nations, the so called *fifth freedom*. As a compromise the Conference developed two separate agreements which were left open for signature. These were the International Air Services Transit Agreement and the International Air Transport Agreement. The International Air Services Transit Agreement provides for what has become known as the first two Freedoms of the Air. It afforded all signatories powers:

* The privilege of flying across the territory of any other signatory power without landing; and,

* The privilege of landing for non-traffic purposes.

The International Air Transport Agreement included the two freedoms of the International Air Services Transit Agreement and added the following three:

* The privilege of putting down passengers, mail and cargo taken on in the territory of the State whose nationality the aircraft possesses;

* The privilege of taking on passengers, mail and cargo destined for the territory of a State whose nationality the aircraft possesses; and,

* The privilege of taking on passengers, mail and cargo destined for the territory of any other contracting State and the privilege of putting down passengers, mail and cargo coming from any such territory (ICAO, 1951).

The International Air Transport Agreement had become known as the *five freedoms*. In reality, the five freedoms are complex, raising difficult issues between States that often have political and economic undertones. This is particularly true of the fifth freedom. For these reasons, only a few of the original signatories remained parties to the International Air Transport Agreement. As a consequence, the exchange of commercial air rights is determined mainly through bilateral negotiations.

ICAO today

> In December, 1994, ICAO was able to look back and celebrate its fiftieth anniversary. Fifty years that were marked by continuous growth, improvements in safety and to having contributed significantly to the social and economic development of States (Kotaite, 1994).

Growing out of the Provisional ICAO (PICAO), which itself was created as a result of the Chicago Conference of 1944, ICAO came into being in April 1947. At the invitation of the Government of Canada, ICAO established its headquarters in Montreal. In 1996, there were 184 ICAO Contracting States.

Standardization

By all accounts, international civil aviation has been enormously successful over the past fifty years bringing extraordinary benefits to mankind that go well beyond the ordinary pleasures of vacationing. To fully appreciate this success, it is important to realize that it was not an accomplishment of the airlines alone. Indeed, it is a testimony to man's ability to cooperate effectively and harmoniously on a worldwide basis.

Airlines are supported by thousands of trained people on the ground who guide and service the aircraft and check on its progress; air traffic controllers who protect and guide it away from other aircraft and toward its destination; meteorologists to keep it abreast of latest weather information and warn it of potentially serious weather conditions; technicians to operate communications and air navigation equipment, cartographers, training experts, mechanics and flight dispatchers. In an afternoon's flight, an aircraft can cross the territories of many nations in which different languages are spoken and different legalities exist. In all of these operations safety must be maintained and no possibility of unfamiliarity or misunderstanding is acceptable. In other words, there must be international standardization and agreement between nations in all of the technical, economic and legal fields so that the air can be used freely, safely and efficiently (ICAO, 1992a). The successful implementation of universal standards has meant having the proper navigation and other facilities strung along the world's air routes, together with highly trained staff to operate and maintain them (ICAO, 1994c).

ICAO provides the machinery for the achievement of international cooperation in the air. As mentioned earlier, the primary way in which ICAO accomplishes this is through the establishment of international Standards and Recommended Practiccs (SARPs) which cover the technical fields of aviation. Figure 2.2 shows the interrelationship between the various annexcs and the technical sections of ICAO.

These SARPs are incorporated into the eighteen Annexes to the Chicago Convention (ICAO, 1994a) and encompass the following specialty areas:

* Annex 1 Personnel Licensing
* Annex 2 Rules of the Air
* Annex 3 Meteorology Service for International Air Navigation
* Annex 4 Aeronautical Charts
* Annex 5 Units of Measurement to be used in Air and Ground Operations
* Annex 6 Operation of Aircraft
* Annex 7 Aircraft Nationality and Registration Marks
* Annex 8 Airworthiness of Aircraft
* Annex 9 Facilitation
* Annex 10 Aeronautical Telecommunications
* Annex 11 Air Traffic Services
* Annex 12 Search and Rescue

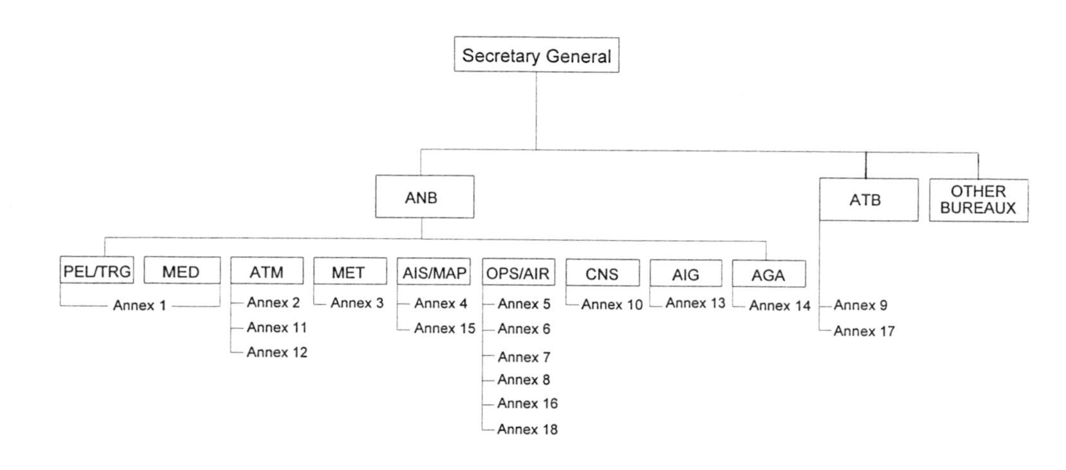

Figure 2.2 ICAO Annexes and their relationship to the various technical sections of ICAO

* Annex 13 Aircraft Accident Investigation
* Annex 14 Aerodromes
* Annex 15 Aeronautical Information Services
* Annex 16 Environmental Protection
* Annex 17 Security
* Annex 18 The Safe Transport of Dangerous Goods by Air

In a relatively short period of time between 1 November and 7 December 1944, the delegates at the Chicago Conference created fifteen Annexes to the ICAO Convention, covering the basic elements of international civil aviation. It is obvious that the delegates had great foresight, however, not quite enough to envision that rules governing environmental protection, aviation security and carriage of dangerous goods would be needed (Ott, 1994). Annexes dealing with these subjects were eventually developed bringing the current total to eighteen. The set of international standards incorporated under these Annexes has become the core set of standards and regulatory material for world aviation. Additionally, the network of facilities, services and procedures so far approved by the ICAO Council and its Air Navigation Commission add up to more than 60,000 items (ICAO, 1994c). A brief description of each Annex is at Appendix A.

The ICAO structure

ICAO has a sovereign body, the Assembly, and a governing body, the Council. The Council, the Air Navigation Commission, the Air Transport Committee, the Legal Committee, the Committee on Joint Support of Air Navigation Services, the Finance Committee, the Committee on Unlawful Interference, the Personnel Committee and the Technical Cooperation Committee, provide the continuing direction of the work of the organization. Corresponding to each committee, division and the Air Navigation Commission, is a section of the ICAO Secretariat, made up of staff members competent in their fields of specialty. The Secretariat is headed by a Secretary General. In order that the work of the Secretariat would reflect a truly international approach, personnel are recruited on a broad geographical basis.

The Assembly The Assembly is considered as the sovereign body of ICAO and is what makes ICAO a global organization of States. It normally meets every three years to review the work of the Organization in detail and to establish the operating budget for ICAO. It is comprised of high level delegates of the member States. The Assembly is also what makes ICAO a truly global organization on civil aviation matters. In the 1990s, the Assembly has been focusing on the preparation of a broad global strategy which could guide States in the planning and implementation of their own civil aviation development (ICAO, 1994c). The work of ICAO is guided mainly by its Contracting States (see Figure 2.3).

The Council The Council is the governing body of ICAO. One of the major duties of the ICAO Council is to adopt international SARPs prior to their incorporation into the Annexes. Once a standard is adopted by the Council, Contracting States have an obligation in accordance with the ICAO Convention, to implement the standard in its territory. As aviation technology advances, the standards are reviewed and amended in order to keep them up to date.

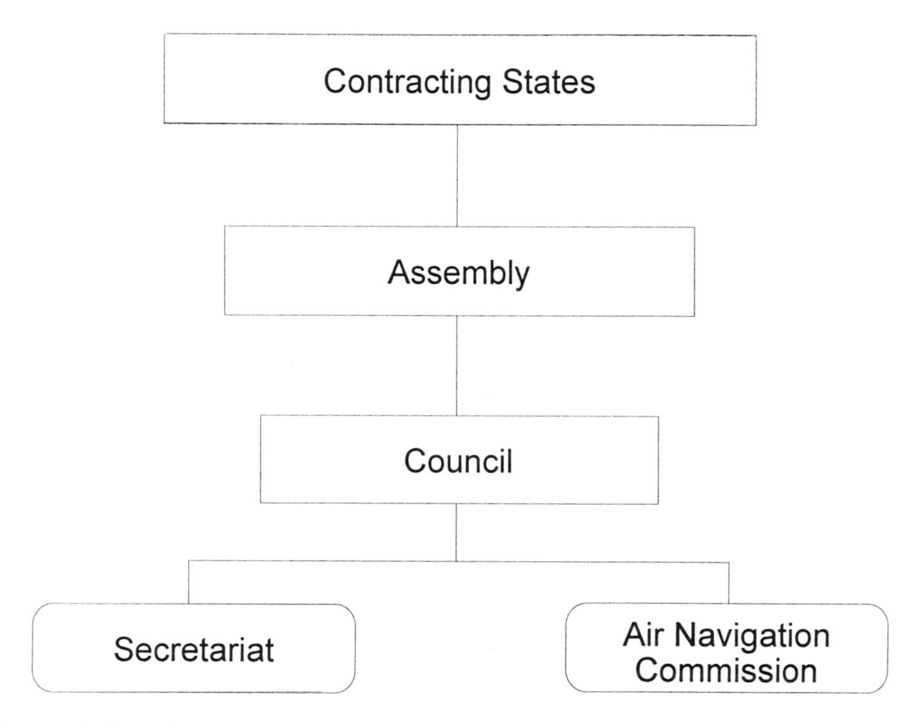

Figure 2.3 A broad overview of the ICAO structure

The Air Navigation Commission Although the Council has the responsibility for adopting the international standards and approving the procedures associated with these, the principal body responsible for the development of these standards and procedures is the ICAO Air Navigation Commission. As provided for in the Convention, and later amended, the Commission is composed of nineteen people of "suitable qualifications and experience in the science and practice of aeronautics" (ICAO, 1994b, p. 22). The Commission reports to the Council and is responsible for the examination, coordination and planning of ICAO's technical work programme in the field of air navigation.

The Secretariat The Council, the Air Navigation Commission and the various committees are assisted in their work by an internationally recruited secretariat. Headed by a Secretary General, it provides the permanent organizational framework for ICAO and provides technical and administrative support to the Contracting States, the Assembly, the Council and its Committees, as well as to the Air Navigation Commission.

The Regional structure Shortly after establishment of ICAO in 1944, the interim Council recognized a need to divide the world into air navigation regions in order to facilitate the planning and implementation of ground services and facilities necessary for international air transport operations. The two principal concepts which led to this conclusion were:

* That the operational and technical problems in different parts of the world vary considerably, therefore, planning and implementation of the required ground services should be carried out on a regional basis; and,

* It is preferable to plan the requirements for air navigation facilities and services through consultation among a limited number of States rather than on a worldwide basis (ICAO, 1993b).

While the Headquarters of ICAO is located in Montreal, Canada, there are seven regional offices which are located in Bangkok, Cairo, Dakar, Lima, Mexico City, Nairobi and Paris. Each is accredited to a group of Contracting States and is charged with assisting those States in the application of the ICAO SARPs as stipulated in the ICAO Annexes, and with implementation of the regional air navigation plans.

Regional planning bodies As part of regional planning activities, the ICAO Council developed regional planning bodies, which are made up of representatives of nations who meet as the need arises and develop and keep up to date, a regional air navigation plan, which is reviewed and approved by the ICAO Council. This plan establishes the requirements for necessary facilities and services for the further development of international civil aviation. Contracting States agree to implement these facilities in line with the ICAO Convention. The regional plan thus serves as a sort of "contract" between the airspace users and the service providers.

There are nine ICAO air navigation regions covering the whole of the earth. The implementation of all of the regional air navigation plans would lead to a globally harmonized air navigation system.

As directed by the Tenth Air Navigation Conference, CNS/ATM systems should be planned and implemented at the regional level through the regional planning groups using the regional air navigation plan as their primary planning document, following coordinated worldwide concepts.

This would lead in time, to a truly global, integrated air navigation system (see Figure 2.4 for a world view of the ICAO regions).

Responsibility and role of ICAO regarding CNS/ATM

> ICAO is at work and will surely develop the technical and institutional framework necessary for the safe and orderly transition to the new technological frontier which will cross sovereign boundaries to establish a seamless air traffic system (Gupta, 1994, p. 60).

In 1983, the ICAO Council gave the task of studying civil applications of satellite systems to the FANS Committee. This allowed aviation specialists drawn from many nations, to sit down together in a global forum and design the system that would meet the needs of the aviation community well into the next century (Campbell and Salewicz, 1995).

In accordance with its obligations under the Chicago Convention, ICAO continues to carry out its responsibility concerning the adoption and amendment of the international SARPs and procedures governing CNS/ATM systems. These SARPs and procedures are continually being reviewed and updated, while new ones are being developed to accommodate CNS/ATM systems. This continuing practice ensures the highest possible degree of uniformity in all matters concerning safety, regularity and efficiency of air navigation. When these SARPs and procedures are incorporated into the ICAO Annexes, the Contracting States of ICAO then have a responsibility under the terms of the Chicago Convention to abide by them or notify ICAO

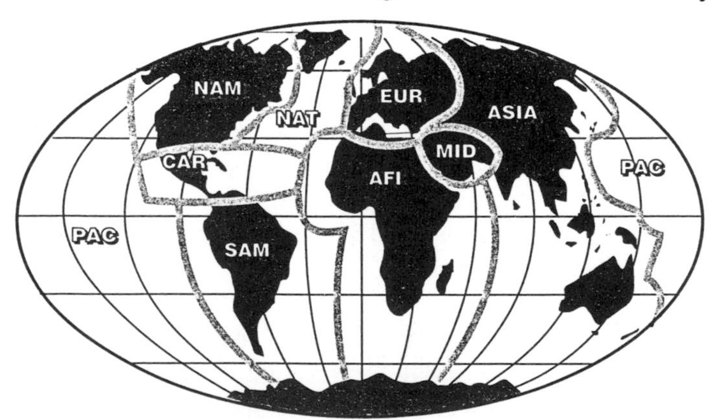

Figure 2.4 The regional structure of ICAO

and thereby the international community of their stated differences. Therefore, decisions that will ensure the continued viability of international civil aviation; decisions that stem from the blueprint provided by the ICAO FANS Committee and driven by advanced technologies, are being made under the worldwide ICAO umbrella.

Furthermore, ICAO is coordinating and monitoring the implementation of CNS/ATM systems on a global basis, through the regional air navigation plans and a global coordinated plan as developed by the FANS Committee. The global plan will be revised and updated in 1997 in order to take into account the latest technology and developments and to assist States in implementing CNS/ATM systems. The global plan will also include an operational concept for ATM which would be supported by the new CNS technologies. ICAO will also continue to provide assistance to States regarding technical, financial, managerial and legal aspects of CNS/ATM systems implementation.

ICAO panels

One of the ways that ICAO accomplishes its objectives is through the development and adoption of international SARPs as mentioned earlier in this chapter. Most of this technical work is carried out by ICAO Panels of the Air Navigation Commission (ANC). ICAO Panels are formed in order to advance solutions to technical problems which cannot be solved adequately or expeditiously by the already established facilities of the ANC or the Secretariat. Panels therefore assist the ANC in its work.

Panels constitute small technical groups of qualified experts, nominated by Contracting States and international organizations and approved by the ANC. Panel members act in their personal expert capacity and not as representatives of any State or organization. The use of panels has evolved and now brings together the best available experts from around the world, under the ICAO umbrella, to examine specialized problems and to find technically feasible solutions which are acceptable to the Contracting States as a whole. The following panels and study groups are at present (1996-7) dealing with CNS/ATM issues:

* Automatic Dependent Surveillance Panel (ADSP)
* Aeronautical Mobile Communications Panel (AMCP)
* Aeronautical Telecommunications Network Panel (ATNP)
* All Weather Operations Panel (AWOP)

* Global Navigation Satellite System Panel (GNSSP)
* Obstacle Clearance Panel (OCP)
* Review of the General Concept of Separation Panel (RGCSP)
* Secondary Surveillance Radar improvements and Collision Avoidance Systems Panel (SICASP)
* Automatic Air Reporting Study Group (ATARSG)
* Flight Safety and Human Factors Study Group (HFSG)
* Frequency Management Study Group (FMSG)
* Operations Study Group (OPSSG)

CNS/ATM Systems Implementation Task Force (CASITAF)

In 1993, the ICAO Council established a high level task force, chaired by the President of the Council and composed of eminently competent people with managerial experience at the most senior executive levels. A wide section of the aviation community was represented on the task force including States, airspace users, service providers, manufacturing, financing and other parties playing a major role in the provision, operation and utilization of CNS/ATM systems. The task force was created to provide advice on the best means for assisting States in the timely and cost-effective implementation of CNS/ATM systems. This task force known as CASITAF has defined several functions that have to do with policy development, technical, operational and regulatory activities.

In November 1994, the Council reviewed the recommendations outlined by CASITAF and agreed on the following principles:

* The ICAO Council is the ultimate mechanism for CNS/ATM systems implementation, however the Council may be advised by a group established for such a purpose. Additionally, a standing entity may be established to oversee CNS/ATM implementation;

* ICAO should review its activities in order to concentrate its resources on CNS/ATM systems implementation. In this way, a CNS/ATM coordination support unit could be established within the existing resources of ICAO;

* Planning and implementation should be coordinated on a regional level which would require the establishment of implementation sub-groups;

* Where possible, States should help each other with implementation and should be encouraged along with other stakeholders to provide staff and other resources to support ICAO (Ruitenberg, 1994).

ICAO CNS/ATM Implementation Committee (CAI)

In light of the results of CASITAF, which had been established by the ICAO Council to advise on how ICAO can best assist States in the timely, cost-effective implementation of CNS/ATM systems, and the finding that ICAO was the ultimate mechanism to perform the monitoring and coordinating function, the Council established the CAI and approved the following terms of reference in February 1995:

* Review progress on the implementation of the CNS/ATM Global Plan, as well as the CNS/ATM implementation plans of States, international organizations, airlines and industry and make proposals to the Council with a view to facilitating the worldwide implementation of CNS/ATM;

* Examine regional problems and requirements, including budgetary, economic and assistance aspects, and advise the Council on ways to ensure inter-regional harmonization and coordination;

* Assess trends and developments related to CNS/ATM implementation and develop recommendations to the Council on possible amendments to the CNS/ATM Global Plan to reflect these matters, and on the impact of these matters on other related work of ICAO; and,

* Refer, as necessary, relevant aspects of its work for comments and advice, via the President of the Council, to any of the standing committees of the Council or to the air navigation Commission.

The CAI began its activities in May 1995 and, on the basis of its terms of reference, adopted a tentative work programme as follows:

* Act as a catalyst to promote the implementation of CNS/ATM systems, using existing resources and structures;

* Monitor and coordinate CNS/ATM related activities both internally and externally to ensure their timely completion and provide guidance to regions and States;

* Assist in securing price and availability of system elements or services;

* Assist in the negotiation for the procurement of systems or services through the development of typical contract clauses;

* Provide the necessary liaison with the International Air Transport Association (IATA) to ensure compatibility in systems and timing;

* Recommend, as necessary, modifications to the global plan to ensure its currency;

* Harmonize the different regional plans into a seamless global ATM system;

* Investigate financing and cost-recovery arrangements; and,

* Assist States in their cost benefit evaluation, selection of technical solutions and implementation options.

In its future work, the CAI intends to establish a consultation process with experts from States, users, manufacturers and service providers, in order to ensure that overall CNS/ATM implementation is always in phase with the realities of the industry and responsive to States and users' needs (ICAO, 1995h).

Other ICAO activities

* Regional planning groups are working on strategies and analyses for their regions;

* The Air Navigation Commission is coordinating technical activities leading to SARPS;

* A global plan for transition to the new system has been prepared and is being updated for completion in 1997;

* The ICAO Council has acted on the recommendations of the Tenth Air Navigation Conference to speed the implementation of CNS/ATM; the Tenth Air Navigation Conference and the 29th ICAO Assembly have endorsed the global system concept for CNS/ATM, and the Legal Bureau has investigated and found "no legal obstacles" to the new systems concept.

Conclusion

> Thus, virtually from aviation's beginnings, the regulatory and operating regimes for civil aviation have reflected either the promise or the reality of advancing aviation.... Today, civil aviation's promise is nothing if not supranational (Freer, 1994).

Since 1944, ICAO has been, and continues to be a highly successful and important international organization. It continues to develop and maintain the eighteen Annexes to the ICAO Convention which cover international aviation's most important technical needs.

Throughout this text, you should now be able to recognize the past, present and continuing role of ICAO and the international community, working through the international legal framework established within ICAO, in development, transition to and implementation of CNS/ATM systems.

A glance at the future

Aviation continues its steady international growth. According to Evans (1994), a United Kingdom Representative to ICAO, "ICAO has proved an effective catalyst to the growth of the industry over the last fifty years" (p. 61). Evans further recognizes the trend toward globalization, and the fact that it is even more important to maintain a forum in the years ahead through which nations can deal with issues of a multilateral nature as they arise. Evans believes that ICAO will continue its contribution in the years ahead to the continuing growth of international aviation to the advantage of all its members.

Recapping the major points

Along with man's earliest experimentation with flight came the recognition by local authorities that some kind of regulation would be necessary. World War I allowed a rapid progression in technology leading to many attempts at international regulation and agreement on a common air law code.

During the course of World War II technological advancements and military operations promoted the aeroplane to a major component of the world's transportation system and a network of air routes and military airfields were established around the world. With the end of World War II in sight, the Government of the United States invited fifty five allied and neutral nations to a conference on civil aviation in Chicago in November 1944. Among other things, the Chicago Conference adopted a new aviation convention which established the International Civil Aviation Organization (ICAO) to foster and guide international civil aviation.

ICAO provides the machinery for the achievement of international cooperation in the air. The primary way in which ICAO accomplishes this is through the establishment of international SARPs which cover the technical fields of aviation. These are incorporated into eighteen Annexes to the Chicago Convention and ICAO Contracting States have an obligation to abide by them or notify ICAO of differences between their own procedures and those of the Annexes.

As the primary international body for the standardization, regularity and harmonization of international civil aviation, ICAO has been the focal point for developing a blueprint for future CNS/ATM systems and will also be the central coordinating body for its implementation. ICAO has fulfilled this role in the following way:

* In 1983, the ICAO Council gave the task of studying civil applications of satellite systems to the FANS Committee which developed an air navigation concept for the future;

* ICAO Panels, which themselves are made up of experts from ICAO Contracting States are carrying out much of the technical work necessary for international standardization;

* The international SARPs are continuously being reviewed and updated, while new ones are being developed;

* These SARPs will eventually be incorporated into the ICAO Annexes and the Contracting States of ICAO have a responsibility to abide by them;

* Regional planning groups are developing the regional requirements for facilities and services to be implemented by individual nations.

Therefore, decisions that will ensure the continued viability of international civil aviation; decisions that stem from the blueprint provided by the FANS Committee and driven by advanced technologies, are being made under the worldwide ICAO umbrella.

The success of international civil aviation and the very fact that CNS/ATM systems could be developed and endorsed by the international civil aviation community at the worldwide level, is a testimony to the success of internationalism in aviation. It is important to understand this international cooperation and its mechanisms to fully appreciate and understand CNS/ATM systems and the outlook for its successful implementation.

Questions and exercises to expand your knowledge

1) Imagine for a minute that the Chicago Conference had not found success, or that the nations signing the Convention had not acted in good faith upon the articles of the Convention. What do you think would have been the overall situation concerning civil aviation today? In your answer consider the following elements:

   * Safety;
   * Cost of travel;
   * Aerospace manufacturing industry;
   * Technology in aviation.

2) What was the urgency for the United States in convening a conference and developing a convention on international civil aviation?

3) As we move toward the future, it seems that the world is becoming more closely tied together. Considering communications, how do you think the use of satellites will play a role in developing a more close knit global community. You can expand your answer beyond the field of civil aviation.

4) Explain how World War II has played a role in advancing the following:

   * Navigation;
   * Surveillance;
   * Air transport.

5) Some people complain that the ICAO "machinery" is too slow. In other words it takes a long time to reach agreements and to accomplish the complex work related to international civil aviation.

   a) explain why there may be a usefulness to this slow method of accomplishing its tasks and why it may even be unavoidable.

   b) explain the disadvantages of the slowness at which ICAO accomplishes its work.

6) Describe an imaginary flight from London to New York. In carrying out this task consider international standardization and its relation to each of the following elements of the civil aviation infrastructure:

   * Air traffic control;
   * Licensing;
   * Meteorology;
   * Navigation;
   * Communications;
   * Aeronautical charts;
   * Environmental protection;
   * Aeronautical information;
   * Search and rescue.

7) It is widely accepted today that ICAO has been a highly successful international organization. How would you justify this? Can you dispute this?

8) Describe ICAO's role in CNS/ATM systems development and implementation.

PART B

COMMUNICATIONS, NAVIGATION, SURVEILLANCE/AIR TRAFFIC MANAGEMENT (CNS/ATM)

Common controller workstation with three positions, including full colour, very high resolution displays for presentation of the traffic environment and flight data information. Voice communications panels are mounted on the console. (Picture provided courtesy of Hughes.)

3. Communications

Introduction

Communication is a vital part of the provision of air traffic services (ATS) and its timely and dependable availability has a significant effect on the quality of the service provided. This is true not only of the availability of communication means, but of the quality of performance and reliability (ICAO, 1984).

The primary objective of an aeronautical communication service is to ensure that telecommunications and radio aids necessary for the safety, regularity and efficiency of air navigation are continuously available and reliable.

As mentioned in previous chapters, CNS/ATM systems will be based mainly on satellite technology. An important application of satellite technology in aviation will be in the provision of communication services for ATM purposes. The integrity and availability of these services must be of the highest order and guaranteed at all times.

As you progress through this text you will recognize that the utilization of communication, navigation and surveillance systems will be vastly different than it is today, primarily because the future CNS systems will make greater use of satellite technology. Furthermore, greater cost efficiencies will be achieved through the transfer of information using digital data techniques along with the gradual implementation of advanced computational methods, all leading to a more effective ATM system.

This chapter provides background on the work of the FANS Committee and identifies the functional shortcomings of present communication systems, the changes that will form the basis of future communications systems and methods, and also reviews the benefits to aviation that could be expected with implementation of these new systems. This review includes a discussion of both data link techniques and the anticipated aeronautical telecommunication network (ATN), as these technologies form the foundation of the CNS/ATM communications element. This chapter also looks at some of the ways and places where the communication component of CNS/ATM technology is being researched and implemented.

Definitions

In order to fully understand this chapter, it is necessary to first become familiar with a few of the internationally recognized terms, used when referring to aeronautical communications and which you will see often as you become involved with international civil aviation.

> *Aeronautical Fixed Service* (AFS) refers to telecommunications between fixed points or ground stations, primarily for aviation, such as between two air traffic control (ATC) units. This is commonly referred to as ground to ground communications.
>
> *Aeronautical Mobile Service* (AMS) refers to telecommunication between aeronautical stations on the ground and aircraft stations. This is commonly referred to as air/ground communications.

Aeronautical mobile communications include:

a) safety communications requiring high integrity and rapid response:

   1) safety related communications carried out by air traffic services for ATC, flight information and alerting service;

   2) aeronautical operational control (AOC) communications carried out by aircraft operators, which also affect air transport safety, regularity and efficiency; and,

b) non safety related communications:

1) correspondence of aircraft operators (aeronautical administrative communications (AAC); and,

2) public correspondence (aeronautical passenger communications (APC).

Aeronautical Telecommunication Network (ATN) An inter-network architecture that allows ground, air-ground, and avionic data sub-networks to inter-operate by adopting common interface services and protocols based on the International Organization for Standardization (ISO) Open Systems Interconnection (OSI) reference model.

Aircraft Earth Station (AES) A mobile earth station in the aeronautical mobile-satellite service located on board an aircraft.

Air Traffic Services Communications (ATSC) refers to communications related to ATS including ATC, aeronautical and meteorology information, position reporting by aircraft and other services related to the safety and regularity of flight (ICAO, 1985b).

Aeronautical Operational Control (AOC) non public communications required by aircraft operators to discharge their obligations under the Annexes to the ICAO Convention in order to exercise authority over the initiation, continuation, diversion or termination of a flight in the interest of safety of the aircraft and the regularity and efficiency of the flight.

Flight Management System (FMS) An integrated system, consisting of both airborne sensor, receiver and computer with both navigation and aircraft performance data bases, which provides performance and RNAV guidance to a display and automatic flight control system.

Ground Earth Station (GES) An earth station in the fixed satellite service, or, in some cases, in the aeronautical mobile-satellite service, located at a specified fixed point on land to provide a feeder link for the aeronautical mobile-satellite service.

Open Systems Interconnection (OSI) reference model A model which provides a standard approach to network design introducing modularity by dividing the complex sets of functions into more manageable, self contained functional layers.

SSR Mode S radar An enhanced mode of secondary surveillance radar (SSR) that permits the selective interrogation of Mode S transponders, the two way exchange of digital data between Mode S ground stations and transponders, and also the interrogation of mode A/C transponders.

Shortcomings of the present system

As mentioned in Chapter 1, the ICAO Council recognized the steady growth of international civil aviation preceding 1983 and, taking into account statistical forecasts at the time, considered the future requirements of the civil aviation community. As an outcome, the Council determined that a thorough analysis and reassessment of the ideas, technologies and institutional arrangements serving civil aviation was in order. In further predicting that the systems and standards supporting civil aviation, because of their inherent limitations, were incapable of dealing with expected growth, the Council tasked the FANS Committee with studying, identifying and assessing new technologies, including satellite technology, and making recommendations for the future development of air navigation for civil aviation over a period of the order of twenty five years.

The FANS I Committee, early in its work, recognized that for an ideal worldwide air navigation system, the ultimate objective would be to provide a cost-effective and efficient system that would permit the flexible employment of all types of operations in as near four dimensional freedom (space and time) as their capability would permit. With this ideal in mind, it was recognized that the existing overall air navigation system and its sub-systems, suffered from a number of shortcomings of a technical, operational, procedural, economic and implementation nature (ICAO, 1985a). After close analyses, FANS I ascertained that the shortcomings of current systems (FANS I conducted its work between 1983 and 1988) around the world amounted to essentially three factors:

* The propagation limitations of current line of sight systems;

* The difficulty, caused by a variety of reasons, to implement current CNS systems and operate them in a consistent manner in large parts of the world;

* The limitations of voice communication and the lack of digital air to ground data interchange systems to support automated systems in the air and on the ground (ICAO, 1988b).

Although the effects of the limitations were not the same for every part of the world, the FANS I Committee foresaw that one or more of these factors inhibited the desired development of ATM almost everywhere. As the limitations were inherent to the existing systems themselves, the FANS Committee realized that there was little likelihood that the global ATS system of the time could be substantially improved unless new approaches were adopted by which the limitations could be surmounted and which would further permit ATS systems to evolve into ATM systems and be more responsive to the needs of the users. CNS/ATM systems therefore, would have to allow for a considerable improvement in the flexibility, efficiency and safety of ATM on a global basis.

Communications

In aviation, the field of communication encompasses a broad range of activities which includes the operation of navigational aids on the ground, in the air and in space and the development of associated technical requirements, specifications and procedures. Communication equipment consists of radars and landing aids as well as air to ground and ground to ground telecommunications equipment.

Telecommunications in aviation can be broadly categorized into either aeronautical fixed service (AFS) or aeronautical mobile service (AMS). AFS comprises communications between fixed points, or primarily between ground stations, such as between two ATS units, or between a meteorological station and an ATC tower. AMS refers to communication between aircraft, or between aircraft and other aeronautical fixed stations, such as an ATS unit.

The bulk of voice communication between ground and air in present systems is carried out using very high frequency (VHF) radio voice for short range communications and high frequency (HF) radio voice for long range communications.

Communications envisaged in an integrated global CNS/ATM environment

Generally, it is expected that in a global CNS/ATM systems environment, a large volume of communication services will be carried out using digital means employing data links, eliminating the need for voice communications in many situations. However, instantaneous voice services are expected to remain available for the foreseeable future, for emergencies and for other communications that may not be suitable for data communications, such as for very busy areas where critical ATC decisions are made and rapidly issued.

The FANS Committee developed and expanded upon the roles of the various existing and unfolding communication concepts and elements in a CNS/ATM environment (ICAO, 1988b) according to the groupings: *terrestrial communication; dependence on data interchange; the role of the very high frequency (VHF) radio band; the role of the high frequency (HF) radio band*; *SSR mode S data interchange*; *and utilization of aeronautical mobile-satellite service bands (AMSS)*. The foreseen roles and evolution of these elements are described below.

Terrestrial communication

Terrestrial, or, earthbound communications will continue to be needed because, from an economic viewpoint, it is far easier to maintain service continuity during frequent aircraft movement in terminal areas, with nearby terrestrial system elements than it is with satellites. Only if service continuity is maintained can safe and efficient ATC be carried out in high traffic density terminal areas. Direct terrestrial communication systems with their short transmission delays are better suited where the rapid exchange, short transaction style of voice communication is required. Therefore, while it is anticipated that there will be a trend towards more data link communications for many functions, voice communication is anticipated to be needed for a long time into the future.

Dependence on data interchange

It is expected that aeronautical mobile communications (air to ground) will make extensive use of digital data interchange between ground and aircraft. This type of service allows data to be communicated between users in the ATM system, eliminating the need for voice communication. It also allows airborne and ground based automated systems to exchange information directly.

The role of the very high frequency (VHF) radio band

VHF is currently used for ATS (voice only) and AOC (voice and data) communications, within line of sight range. The excellent operational reliability and the number of channels available, make VHF the basis for safety communications in many continental areas. Saturation of VHF channels, however, is already becoming a reality. Because of its propagation characteristics, the use of VHF will be limited to line of sight communication and therefore worldwide coverage is not possible.

In addition to voice, VHF will be used for data link communications. You will read more about data link later in the chapter.

The role of the high frequency (HF) radio band

High frequency communication is the only means available for "over the horizon" contacts (e.g., over the oceans and remote areas where VHF is not available). HF communications have reliability limitations imposed partly by the variability of propagation characteristics. Aeronautical operations are currently limited by the range of possibilities offered by HF. The replacement of HF voice by satellite communication will be a major step forward. This replacement will be delayed in the more extreme polar regions, for economic reasons. Aircraft wishing to fly over the areas north or south of available satellite communication coverage could either carry HF radios or choose to carry additional fuel to compensate for the increased track distance required by a flight plan that avoids areas for which there is no satellite coverage.

In addition to the areas mentioned above, extensive experimentation is taking place using HF for data link communications.

SSR mode S data interchange

SSR mode S radar is an enhanced mode of SSR that permits the selective interrogation of Mode S transponders. Many of the surveillance related improvements associated with Mode S are based on its ability to uniquely address each aircraft within its range. Additionally, Mode S will permit the two way exchange of digital data between mode S ground stations and transponders, and also the interrogation of mode A/C transponders. Mode S radar will therefore provide an air-ground data link. Because Mode S is primarily a surveillance radar, it is discussed in more depth in Chapter 5. By application of an OSI concept (see definition above and explanation below), Mode S data link will be interoperable with satellite or aeronautical mobile satellite service (AMSS) data link. SSR mode S data link will be used for ATM in terminal areas and some other high density traffic airspaces.

Utilization of aeronautical mobile-satellite service bands (AMSS)

The exploitation of satellite technology in the future global air navigation infrastructure is the key to CNS/ATM systems development that will have benefits for international civil aviation into the next century. The FANS Committee developed a system architecture for satellite communication services that provides for all four kinds of communication (i.e., ATC, AOC, AAC and APC) (see definitions at the beginning of the chapter), and encourages multiple user participation through a minimum system capability up to a complex multi function system. This architecture permits integration of services to the degree that institutional considerations, operational priorities and the need to preserve frequency spectrum for aeronautical safety communications may allow. The architecture enables both data and voice communication. It provides for flexibility in the design of systems and their evolutionary development in the future. Also, by applying the OSI concept, the architecture offers interoperability of the satellite elements of the system with the aircraft and with ground communication networks, including interoperability of satellite services of separate providers. This system architecture has been approved by the ICAO Council and is the basis for the development of International SARPs (see Chapter 2 and Appendix A for a description of SARPs) on satellite air-ground communication systems by the AMC Panel (see Chapter 2 for a general description of ICAO panels).

It is not expected that satellites will meet the requirements of all airspaces, particularly in terminal areas or in other high density areas where direct ground to air communications would be preferable. Figure 3.1 depicts a communications system overview taking advantage of all of the available elements of the future system.

Data link

> If a flight management system (FMS) is in place (which will be the case in most commercial airline and business aircraft), the preprogrammed intended trajectory of the aircraft will also be known. What is known to the aircraft system can be made known to the ground system.... (Andresen, 1993, p. 22).

The employment of data link in CNS/ATM systems will enable computer to computer digital communications and can therefore accommodate the transfer of any digital messages between two end users. Data link is one of the critical elements of CNS/ATM technology and it is therefore necessary to acquire an understanding of the fundamentals of data link.

- *The Secondary Surveillance Radar Improvements and Collision Avoidance (SICAS) Panel*

The SICAS Panel was formed by the ICAO Air Navigation Commission in order to undertake specific studies with a view toward developing draft International SARPs (see Chapter 2 for a general description of ICAO panels and SARPs), procedures and, where appropriate, suitable guidance material concerning SSR enhancements and related data link and collision avoidance systems.

In 1989, the Air Navigation Commission expanded the role of the SICAS Panel to include the development of ICAO material to aid with the formation of systems which would have maximum commonality and interoperability between ATS data links, including satellite data links. This task emerged from the work of the FANS Committee which emphasized the need for the interchange of digital data over dissimilar aeronautical data links. The FANS Committee also recommended that the principles of the ISO - OSI

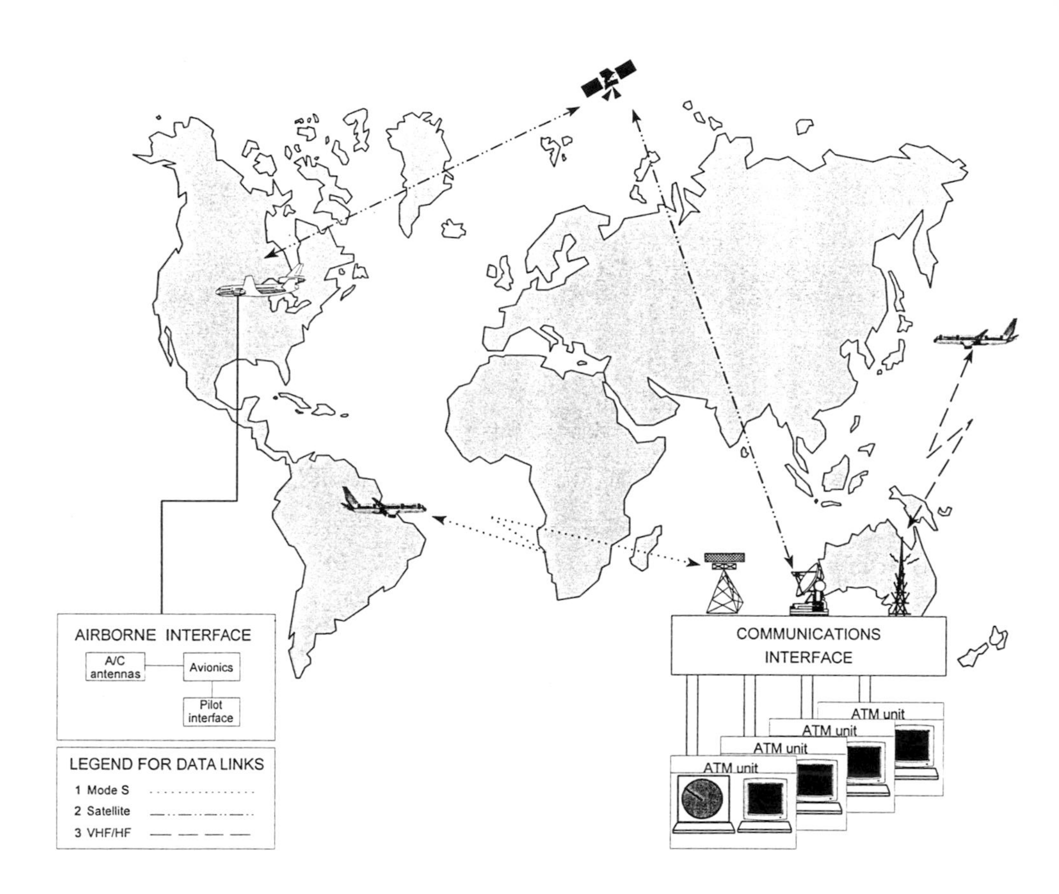

Figure 3.1 A future communications system taking advantage of all of the available elements

architecture be applied in developing aeronautical data links in order to provide for their interoperability.

Subsequent studies undertaken by the SICAS Panel resulted in the concept of the ATN. At its fourth meeting, the SICAS Panel developed a description of the ATN which was then published for worldwide use in an ICAO manual (the ATN is described in more detail later in the chapter).

An ATN Panel was established in order to develop the international SARPs, procedures and guidance material under which the ATN would be implemented.

ATS applications of data link

The SICAS Panel (ICAO, 1988c) identified initial ATS surveillance applications and benefits which would accrue from the transfer to the ground, using data link, of aircraft information such as identification, altitude, status and trajectory information. Although the initial work of the SICAS Panel was concerned mostly with SSR mode S data link, the same benefits could be roughly envisaged for other types of data link applications, such as HF, VHF and satellite.

In addition to the above, the SICAS Panel recognized that flight information service, which includes the dissemination of essential weather and traffic information, could be enhanced resulting in reduced radio transmitter occupancy times, reduced request/response times, a decrease in controller workload, more timely dissemination of information and a more comprehensive service than even the most experienced controllers are capable of providing.

The SICAS Panel concluded that the overall ATS service and the provision of flight information service could be enhanced using data link techniques as follows:

Weather information The data link will enable pilots to have direct access to meteorological data from ground based systems. Warnings associated with severe weather conditions such as windshear could be quickly delivered to pilots and measurements made on the aircraft could be transferred to the ground, either on request or on a regular basis.

With the increasing importance being placed on the use of flight path prediction techniques, improvements in the quality of the available wind and temperature data will be necessary before this technology can be fully exploited. The equipment carried on board many of today's commercial aircraft is capable of accurately measuring wind vector and air temperature.

Aeronautical information All types of aeronautical information may be provided to pilots on request, and could include information such as that contained in automatic terminal information service (ATIS) broadcasts, information on changes in the usability of navigational aids and information on airspace class and restrictions.

Airspace information Airspace information transmitted via the data link may include minimum safe altitude warnings and airspeed restrictions. It could also automatically provide warnings to pilots concerning predicted incursions into control or restricted areas when this would happen without a proper clearance.

Navigational information Modern radar data processing systems may be regarded as dynamic data bases in which present and predicted aircraft positions as well as related data are stored. Navigational information could be derived from this data and reported to an aircraft via the data link on pilot request.

Conclusions on data link

The SICAS Panel noted the evolutionary nature of data link implementation which, at first must be designed so that controller workload is not increased in a mixed data link environment, that is, a mixture of data link and non data link equipped aircraft.

Furthermore, the successful utilization of data link depends to a large extent on the degree of automation available to the controller and the pilot. There have been a number of studies on the effects of providing controllers with data link capabilities without also offering ATC automation. Generally, such efforts fail to significantly reduce controller workload because it is more natural and efficient to communicate by voice than it is to input data via a keyboard (ICAO, 1988c).

As a result of the above studies, the SICAS Panel recognized that it would not be likely that data link alone would significantly reduce the workload of controllers until there were significant improvements in data entry techniques that would themselves use automated means, such as menu based input systems, and/or reliable voice recognition systems connected to automated message formatting systems. More importantly, the availability of data link and automation would make it possible for aircraft to interact directly with ground data bases which would enable the extraction of weather data from ground systems and other safety related routine and non-routine information.

In the early stages, an automated ATM system could provide additional information to assist the controller in conflict prediction and prevention. This information could be provided from aircraft down linked data, such as next position and estimated time over that position. Aircraft crews would also be able to receive automatic clearance confirmations, after they have been implemented by controllers, and pre-notifications, such as when aircraft should change frequencies.

In later stages, the ATC system may evolve such that advice to manage air traffic will be generated automatically. At first, this information could be approved by the controller beforehand. Later, however, automatic interchange could conceivably take place between flight management systems (FMS) and the ATC system in order to exploit three and four dimensional navigational capability as a means of establishing conflict free trajectory planning.

Overall, data link will allow the direct air/ground - machine to machine communication which will enhance further ATC automation, necessary to provide the relevant automated functions in CNS/ATM systems and also to exchange information between flight management and ATM systems. These capabilities are essential to increase airspace capacity and improve productivity (ICAO, 1988c) and to realize the full benefits of CNS/ATM.

Aeronautical Telecommunications Network (ATN)

> It is described as the glue binding together the future air navigation system, a common communications architecture, a telephone exchange for data, and a communications backbone (Daly, 1994a, p. 29).

An integrated global CNS/ATM system will make increasing use of data transfer between airborne and ground based systems. This data will encompass at first, basic ATC information that is now carried out using voice, such as the issuance of clearances and weather information.

As automation and computational methods are increasingly employed, there will be a need for digital data transfer between airborne and ground based automated systems. Considering the increasing availability of satellites for communications, the emergence of mode S and the existence of VHF and HF, a concept has been developed which will enable rapid and reliable aeronautical communication between air and ground, almost anywhere on the globe, *interfacing* the communication services of different user groups.

The ATN can be paralleled with the Internet: the electronic network which links subscribers the world over. ICAO has defined a similar network to the Internet for civil aviation. An internet is a shared network of many nodes, or addresses. Any subscriber can transfer any kind of data to any other address, or a number of addresses, on the system. In the same way, ATN nodes can be physically located anywhere: at an ATC Centre, on an aircraft, at the gate or in a hangar. Remember also that many of the ATN nodes are mobile, moving at great speeds.

The key element and access point to the ATN is called a router. The job of a router, which will usually be a software function running on a multipurpose communications device, is to ensure that data messages (called packets) are properly addressed to their destinations elsewhere in the ATN, and to receive ATN messages addressed to the terminal or network that it serves. Figure 3.2 displays the ATN architecture including the routers.

ATN is not dependent on airborne satellite communications. Over land, or within line-of-sight distance of the coast, ATN messages can be transmitted by VHF digital radio or Mode S radar (Sweetman, 1994).

In a more technical sense, the ATN is a set of communications protocols based on an OSI (OSI is explained in more detail later in the chapter). OSI will allow the interconnection of airborne, ground and air/ground networks, including Local Area Networks (LANs), Wide Area Networks (WANs) and radio links (RF) allowing a seamless, transparent, worldwide aeronautical digital data communications network (Ryals, 1993). The ATN, therefore, through the use of an agreed communication protocol structure, will provide for the interchange of digital data packets between end users of dissimilar air-ground and ground-ground communication sub-networks (e.g., SSR Mode S, VHF, HF, AMSS), resulting in a common data transfer service.

Briefly, a position report by an aircraft over oceanic airspace for example, could be transmitted using a satellite carrier. As the flight came within VHF coverage, data could be transmitted over VHF data link and, when within range of mode S radar, data could be transmitted via mode S data link, thereby relieving a congested VHF band. In the context of the ATN, all of this would be transparent to the end user, who would have no knowledge, nor any need for it, of the path taken by the data (Cole, 1990). The selection would be made automatically on the basis of channel availability, the type of data, the tolerable delay between question and answer, etc. It will be a truly open system, capable of many interconnections.

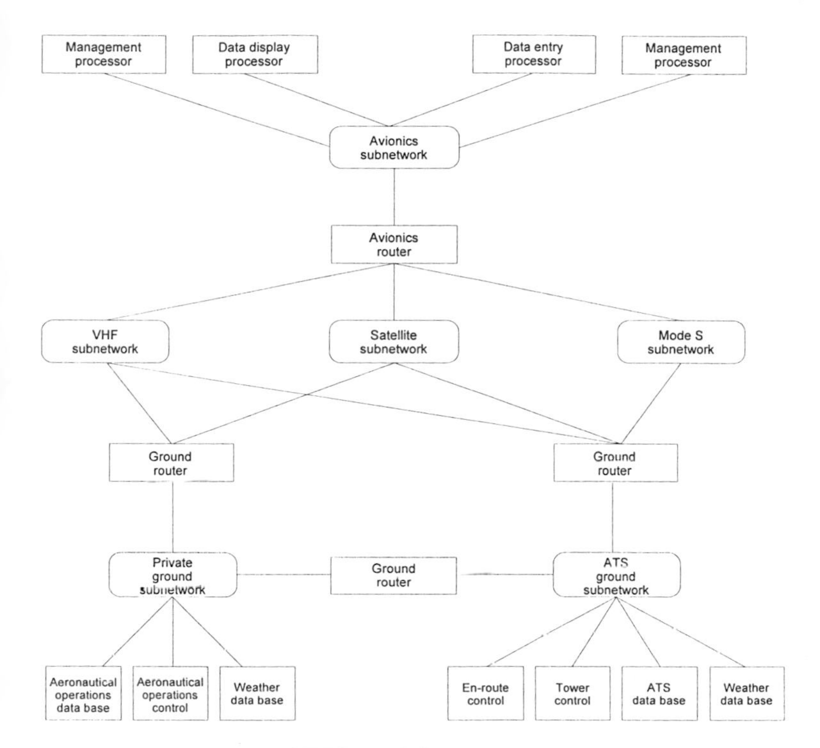

Figure 3.2 The ATN architecture

Global data networking

In order to fully utilize automated ATM systems, a global data networking infrastructure must be implemented which supports the inter-networking of state of the art computer systems operating in fixed ground based and mobile aircraft based locations (ICAO, 1991b). The key to success in developing and implementing this new inter-network infrastructure is the recognition that:

* Increased use of distributed ATM automation requires an increased level of computer to computer data communication, including *data communication between aircraft based and ground based computers serving fixed and mobile users;*

* Increased levels of distributed ATM automation require a more richly connected and *more flexibly configured data network*

infrastructure than exists today, both in aircraft based and ground based environments; and,

* Real success in ATM automation can only be achieved when *aircraft based computer systems are designed and implemented as data processing and networking peers to their respective ground based computers,* rather than continuing in their current role as aircraft terminals attached to ground based hosts (ICAO 1991b) (see Figure 3.3).

Open systems interconnection (OSI)

The ATN design is an inter-network architecture that will allow ground, air-ground and avionic data sub-networks to inter-operate and share data by adopting common interface services and protocols based on the ISO - OSI reference model.

This allows communication services for different user groups. The design provides for the incorporation of different air-ground sub-networks (e.g., SSR Mode S, AMSS) and different ground-ground sub-networks, resulting in a common data transfer service. These two aspects are the basis for interoperability of the ATN and will provide a reliable data transfer service for all uses (i.e., ATM, AOC, AAC and APC). Furthermore, the design is such that user communication services can be introduced in an evolutionary manner.

The FANS Committee has long recognized the virtues of adopting the OSI model of data communications and its implementation within the aeronautical community as the ATN. The benefits of adoption of an OSI based communication system such as the ATN include:

* Interoperability between different computer systems and peripherals;
* Exchange of information on a global basis; and,
* Freedom to choose innovative products (ICAO, 1993a).

Service providers

With all the discussion in this text of satellites and communications, it is appropriate to mention that it is all but impossible for each and every State

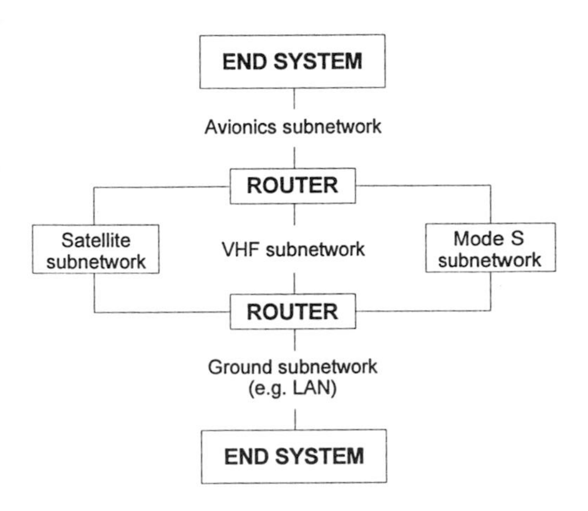

Figure 3.3 ATN components

to put into space its own satellite network to provide the necessary services associated with CNS/ATM. Furthermore, the infrastructure required to support the flow of communications across borders and through space could not be feasibly supported by any one State or organization. In light of the difficulties involved, several international organizations and corporations that provide satellite and communication services to the international aviation community through specialized arrangements have been established. Three of the most prominent of these service providers (i.e., the International Maritime Satellite Organization (Inmarsat), Aeronautical Radio Incorporated (ARINC), the International Society for Aeronautical Telecommunications (SITA)) are described below so that you can gain an awareness of the scope of cooperation already required for present systems and increasingly necessary for the CNS/ATM systems of the future.

- *The International Maritime Satellite Organization (Inmarsat)*

As you read through succeeding chapters you will see that all aspects of CNS/ATM require the provision of satellite services. In a few cases the satellites are provided exclusively by one or two states as is the case with the GPS of the United States and the Global Orbiting Navigation Satellite System (GLONASS) of the Russian Federation which you will read more about in the next chapter. In most other cases, however, satellite services will have to be arranged through various sharing schemes and marketing agreements.

Presently, the most accessible and convenient worldwide international satellite structure already in place and capable of meeting the needs of the international civil aviation community is that of Inmarsat.

Inmarsat is an international and inter-government organization which was established in 1979 for the purpose of providing the space segment (satellite constellation) necessary for improving maritime, aeronautical and land mobile communications. Inmarsat is required to serve all geographical areas where there is a need for mobile satellite communications and to act exclusively for peaceful purposes. The parties to the Inmarsat Convention are the member States. The signatories to the Inmarsat Operating Agreement are the public or private entities designated by the member States to participate in the financing and management of Inmarsat. Signatories are typically the owners and operators of the satellite ground stations which provide connectivity through the space segment between mobile terminals and ground telecommunications networks (Featherstone, 1993) (see Figure 3.4).

Inmarsat foresaw the applicability of its technology to civil aviation and in 1983 decided, after consultation with ICAO, to expand the maritime mobile spectrum of the transponders planned for its Inmarsat-2 constellation of satellites to include a portion of the spectrum allocated by the International Telecommunication Union (ITU) to aeronautical mobile satellite route services (AMS(R)S).

In the case of communications, which was the first realized potential of its services, Inmarsat had, by the end of 1990 after working closely with ICAO for some years, implemented a worldwide aeronautical satellite communication system.

Featherstone (1993) points out that the economies of mobile satellite communications are such that a satellite constellation dedicated solely to aeronautical applications could not be justified and sharing the space segment with other users was therefore unavoidable. There is, however, a measure of independence provided for communications concerned with the safety and regularity of flight operations. In any case, the Inmarsat Organization and its constellation of satellites came to be seen as an ideal means of early implementation of CNS/ATM technology at a reasonable cost to States.

Anticipating that many States would conclude that the most cost-effective approach to satellite based CNS systems implementation would be to purchase satellite communications for use in ATM, a number of consortia of specialist service providers and Inmarsat signatories had established a

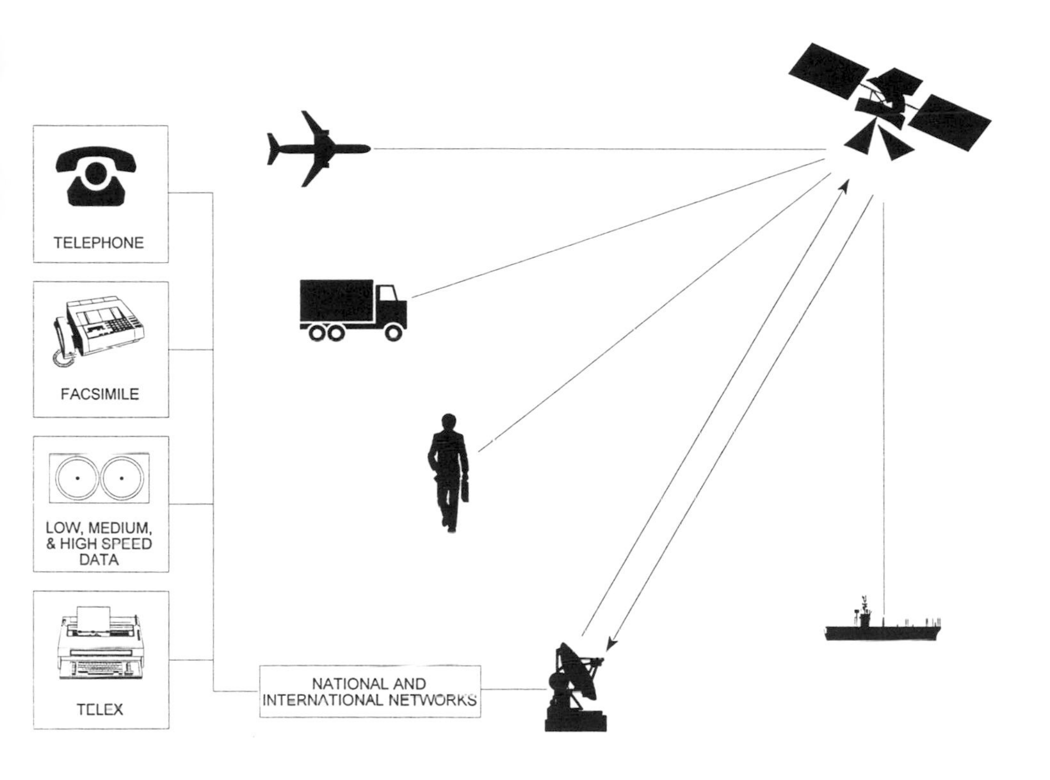

Figure 3.4 The Inmarsat system

market place for such services. They typically offer connectivity from any terrestrial location to aircraft anywhere in the world via their own ground networks and satellite communications ground earth stations in each satellite coverage region. This fact effectively negates the need for every State to own and operate its own ground earth stations.

Inmarsat's satellite capabilities are continually being extended as requirements evolve, especially in terms of service provision to required levels of integrity, reliability and availability. In February 1991, Inmarsat awarded a contract for the manufacture of four Inmarsat-3 satellites. Following a decision taken at the Inmarsat Council Meeting in 1994, the organization ordered a fifth Inmarsat-3 satellite. The first Inmarsat-3 satellite was scheduled to be in operation by March 1996. Successive spacecraft will be launched at four month intervals. The launch order for the satellites is expected to be for the Indian Ocean Region (IOR), followed by the Atlantic Ocean Region-East (AOR-E), Pacific Ocean Region (POR) and Atlantic Ocean Region-West (AOR-W) (ICAO, 1995e). The overall use of these satellites for international civil aviation is being carefully developed through close coordination with ICAO and its bodies.

- *Aeronautical Radio Incorporated (ARINC)*

ARINC is owned by the scheduled airlines of the world and other participants in international civil aviation and provides aviation with communications and information services on a not-for-profit basis. ARINC is governed by a board of directors elected annually by its multinational shareholders. ARINC's network of subscribers, which are deemed to be members of ARINC, is broadly international in character, as are its widespread service operations.

ARINC has been involved in many aspects of civil aviation related to communications, including frequency spectrum management, since its establishment in 1929. Additionally, ARINC often represents aviation's interests at many international fora when issues associated with communications are being addressed.

In addition to the above, ARINC also operates an Airline Communications Addressing and Reporting (ACARS) data link system. ARINC designed, developed, deployed and continues to operate the ACARS. This system is utilized by approximately 4,521 data link equipped aircraft and presently handles close to 300,000 data link messages per day. ACARS supports an air-ground link over VHF radio frequencies using a character-oriented message format. Messages are formatted by the ACARS management unit based on data either manually input by the flight crew or automatically generated by other sensors and avionics on the aircraft. The messages are then transmitted, using VHF data link frequencies, through VHF transceivers to ACARS ground stations, which then pass the messages over the ARINC Data Network Service (ADNS) to ACARS processors. The processors then translate the messages to the appropriate format and route them to their intended destinations. These destinations are often the flight operations centres and are for operational control. Increasingly, however, the ACARS network is being used to send and receive ATS messages.

In 1992, United Airlines began using its ACARS avionics over the Pacific Ocean to communicate ATS data over the Inmarsat satellite network. Trials and demonstrations have continued since that time and are now being used for ATC operational purposes in many areas of the world for ATS purposes. This aspect of the ACARS is further discussed later in the chapter, under the section: A Glance at the Future.

ARINC's engineering staff and its Airline Electronic Engineering Committee (AEEC) are also providing substantial support toward the development of data link protocols, avionics architecture and air/ground networks which are expected to be fully compatible with the planned ATN. In addition to its ACARS network, ARINC communication facilities and services will form an integral part of the envisaged ATN.

ARINC has provided extensive technical support to ICAO for many years, including participation on several of its panels.

- *International Society for Aeronautical Telecommunications (SITA)*

SITA is the operator of a data communications network similar to ARINC's described above. SITA's network covers over 200 countries and territories. Between them, the data communications networks of SITA and ARINC cover virtually the entire globe. In addition to its communications network, SITA also provides a wide range of value added network services and information processing services to the air transport industry. SITA, like ARINC, is working actively towards the implementation of the ATN and migration of the existing data link services to the new ICAO standards, which should enable States to benefit from FANS technology at minimum cost and without delay.

Summary of the benefits of the future global communication system

* Linkages between ground and airborne systems will be more direct and efficient, resulting in improved ATM services;

* Handling and transfer of data among operators, aircraft, and ATM providers will be improved;

* Channel congestion will be reduced;

* Errors due to faulty communications will be reduced, and safety thereby enhanced;

* Interoperability across applications will be possible, with minimum avionics required;

* Reduced costs and improved efficiencies will result from the ICAO standardized ATN;

* Existing ground-based systems such as HF voice can gradually be phased out. Figure 3.5 shows the differences between the present and future communications system.

Recapping the major points

As we move toward the future, the utilisation of new communications, navigation and surveillance elements based on the global concept, as well as their interaction with each other and with the ground and airborne systems they are meant to serve, will be vastly different from what they are today. This is due primarily to the fact that CNS/ATM systems will make greater use of satellite technology. Additionally, the transfer of information using digital data techniques, along with the gradual implementation of advanced computational methods, will form an ATM system that will be the backbone of the air navigation infrastructure, allowing greater cost effectiveness and efficiencies than are currently possible.

This chapter provided background on the work of the FANS Committee relating to communications, which included the identification of the functional shortcomings of present communications systems, a description of the envisioned changes that will form the basis of the future system and methods, and a review of the benefits to aviation that could be expected with implementation of the new system.

When reviewing the field of communications, the main features of an integrated global CNS/ATM system, to be implemented within a period in the order of twenty five years, can be highlighted as follows (ICAO, 1993a):

* Satellite data and voice communication services will remain available for the larger part of the world;

* HF will have to be maintained over polar regions and other remote areas and will also be used as a data link medium;

* In the future, air-ground communication will make extensive use of digital modulation techniques to permit high efficiency information flow, optimum use of automation both in the aircraft and on the ground, and economic frequency spectrum utilization;

* Except for high density areas within coverage of terrestrial based communication systems, aeronautical mobile communication services (data and voice) will use satellite-relay;

* Terrestrial based air-ground communications will continue to be used in terminal areas and in other high density airspace;

* VHF will remain in use for voice and certain data communications in many continental and terminal areas;

* The SSR Mode S will provide an air-ground data link which will be used for ATM purposes in high density airspace;

* Interoperability with other data links will be facilitated through the application of the OSI model;

* The ATN, through the use of an agreed communication protocol structure, will provide for the interchange of digital data packets between end users of dissimilar air-ground and ground-ground communication sub-networks.

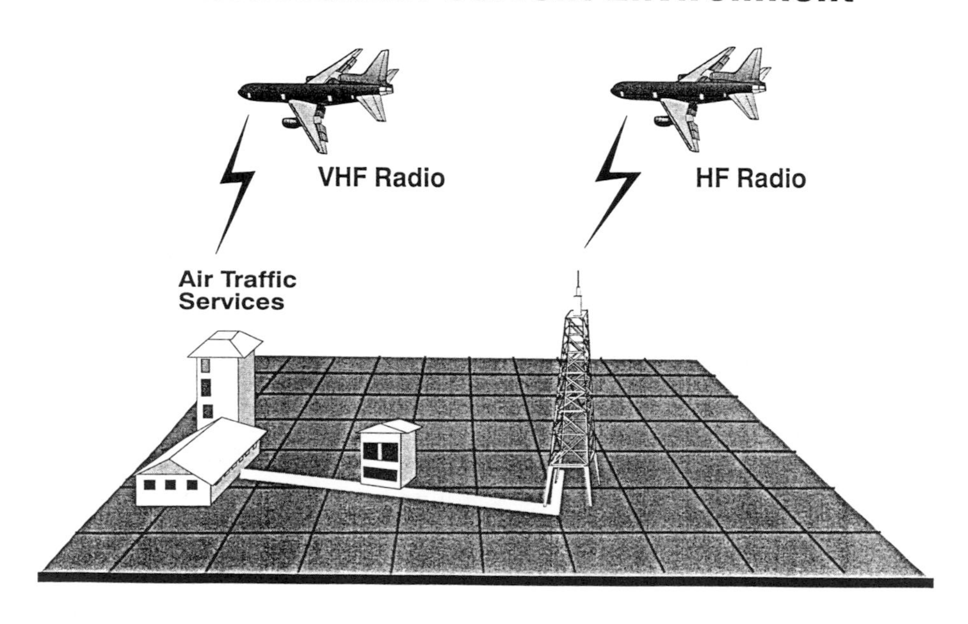

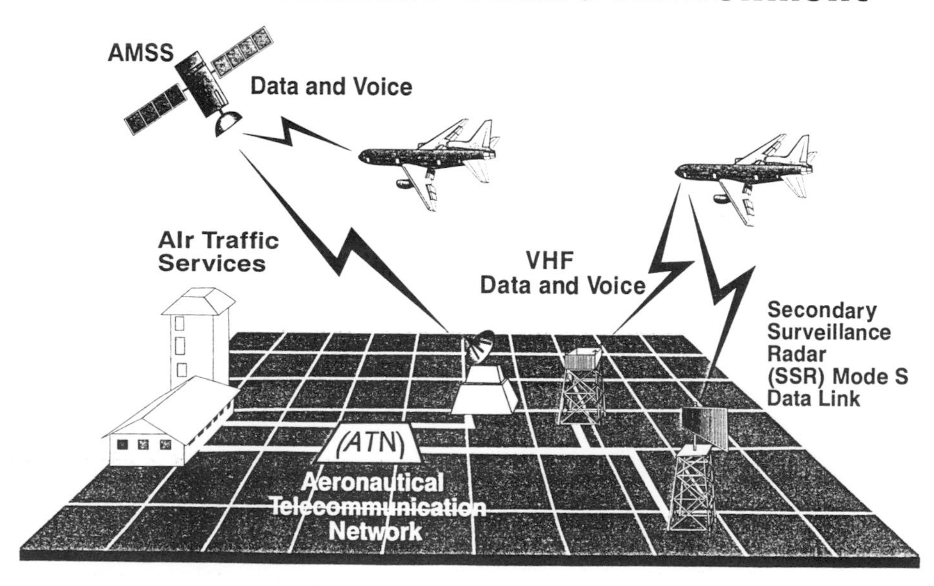

Figure 3.5 The differences between the present and future communications environment

A glance at the future

FANS upgrade packages

The three main manufacturers of aircraft: Airbus, Boeing and McDonnell Douglas are all working, in cooperation with the airlines and avionics manufacturers toward development of systems to bring early benefits of CNS/ATM technology to the airlines for use over the oceans and other remote areas. Immediate benefits are expected as follows:

* Reduced separation between aircraft over oceanic areas, thereby increasing capacity;

* Possibility of using *preferred routings* and *flexible tracks*;

* Quicker response by ATC to altitude change requests;

* Avoidance of altitude loss at points where tracks cross;

* At first, non-precision approaches, based on Global Navigation Satellite Systems (GNSS), and eventually, precision approaches based on GNSS (GNSS is covered in more detail in Chapter 4);

* More availability of diversion airfields because of improved autonomous aircraft capability (Norris, 1994b).

Boeing's FANS 1 package

The primary certification of the FANS 1 package accomplished by Boeing and Qantas airlines, took place in June 1995, followed by the first truly operational scheduled passenger flight across the Pacific by a Boeing 747-400 FANS equipped aircraft. This is an industry milestone, "which some argue is as significant as the introduction of the jet engine or airborne radio" (Norris 1994a). Like the packages planned by its competitors, Airbus and McDonnell Douglas, Boeing's FANS 1 package complements ICAO's global plan.

In its drive to attain early benefits of CNS/ATM technologies, the International Air Transport Association (IATA) persistently raises CNS/ATM issues at aviation meetings, seminars, workshops and other international gatherings and has developed its own "User Driven Plan" for early implementation of CNS/ATM for all parts of the globe. At a gathering in March of 1995 of an ICAO ATS route planning meeting for the Eastern part of the European region, a brief summary of the evolution of the Boeing FANS 1 system was presented in a Working Paper (1995a) by IATA. In their presentation, IATA explained that Boeing initially committed itself to providing avionics functionality on the Boeing 747-400 aircraft so that it would be able to operate in a full CNS/ATM environment. This functionality was originally planned to be achieved through the offering of a modification of the aircraft's Flight Management System (FMS). This upgrade was expected to provide Automatic Dependent Surveillance [ADS (ADS is described in Chapter 5)], Controller-Pilot Data Link Communications (CPDLC) to be used over the planned ATN, and also the integration of GPS into the aircraft's FMS along with many other operational improvements. Boeing withdrew this offer in 1992 because there was little prospect of the ATN being developed and becoming operational in the near term, the specifications for CPDLC were as then undefined, GPS certification requirements were not known and, in the absence of compatible ground systems there could be no operational benefits to provide the necessary incentives for the airlines to purchase the proposed package, nor could aircraft systems be certified without a corresponding and operational ground system.

Despite this initial setback, the aeronautical community in certain areas of the world, in the face of growing airspace congestion, recognized that significant benefits and operational efficiency would only be available from early implementation of new CNS/ATM systems.

A core group of airlines: Air New Zealand, Cathay Pacific Airways, Qantas and United Airlines determined that significant reductions in their operating costs were achievable by flying optimised flexible tracks on their Trans-Pacific routes.

Based on the above, the airlines determined that they needed an avionics upgrade package that would provide the necessary CNS capabilities to support the daily use of a Dynamic Air Route Planning System (DARPS) on their Trans-Pacific routes. This core group of airlines approached Boeing in June 1993 and presented their requirements for an upgrade to the Boeing 747-400 which was named FANS 1.

Work on the project commenced with a development team consisting of:

* The core group of airlines;

* Boeing and its avionics suppliers, such as Honeywell;

* The aircraft communications, addressing and reporting system (ACARS) service providers; and,

* The ATS providers in the South Pacific, Members of the Informal South Pacific ATS Coordinating Group (ISPACG); which consists of the United States Federal Aviation Administration (FAA); the Civil Aviation Authorities of Australia and Fiji; the Airways Corporation of New Zealand and the French Polynesian aviation authorities.

It was agreed at an early stage by this group that a single step replacement of existing systems with fully ATN compliant systems was not acceptable, because of:

* Substantial investments in existing systems and limited resources which drove the need to use existing air/ground communications systems with minimal modification, while ensuring that current and future communications networks coexist during the transition to ATN;

* Little prospect of ATN capability being realised for operational use until later in the decade; and,

* The limited memory capacity and computer throughput of existing avionics equipment which was not capable of incorporating the complex protocols associated with the ATN.

Development strategy

A strategy was therefore developed which permitted the use of existing communications infrastructure and which also enabled the support of bit-oriented applications over existing character-oriented networks with minimal impact on existing avionics. The baseline requirements developed for FANS 1 emphasised that:

* The implementation of CNS/ATM capability must proceed incrementally with benefits for the operators and ATS service providers being readily realisable in the foreseeable future with benefits outweighing costs at each step;

* Plans to implement CNS/ATM functions must consider all of the requirements necessary to achieve maximum benefits both on the ground and in the air;

* There should be maximum utilization of existing equipment and minimum requirement for new equipment or software;

* Integrity issues must be addressed at the system level; and

* The Boeing 747-400 upgrade plans must be consistent with plans for upgrading other Boeing aircraft.

The avionics changes that were necessary to achieve the proposed benefits were the provision of:

* Automatic Dependent Surveillance (ADS);

* Controller-Pilot Data Link Communications (CPDLC);

* Global Positioning System (GPS) integration; and,

* Modifications to make use of ACARS systems (explained below) so that they could support communications between certain ATM applications via existing character-oriented communications networks. Figure 3.6 gives an overview of the FANS 1 functionality.

The aircraft communications, addressing and reporting system (ACARS) As can be seen from the discussion of the ATN earlier in this chapter, the aviation community has long recognized the need for a seamless, standards-based, inter-operable communications infrastructure. The term Aeronautical

Telecommunications Network was in fact coined to identify the resulting, desired networked environment.

The immediate challenge comes from the fact that implementation of data link systems fully compliant with the ATN may take several years and that full deployment of the ATN will be a complex process. In the meantime however, it has been widely accepted by those embracing the FANS 1 concept, that an interim solution is available in the way of ACARS.

It is useful to have some background information on ACARS in order to fully understand the FANS 1 package. ACARS supports the transfer of messages for Aeronautical Operational Control (AOC) and Aeronautical Administrative Control (AAC). In other words, it is used by the airlines for their internal purposes, mainly to communicate with aircraft for AOC and AAC purposes. Therefore, it is felt by supporters that the networking technology which already exists in the form of ACARS, can be put to use as an intermediate measure for air traffic services, by using equipment which the airlines already own. It should be remembered however that ACARS is a *character*-based networking environment, while the ATN would be a bit-oriented standards-based networking environment. Remember also that ACARS was not initially developed to support ATS or ATM functions, which is where a bit of controversy enters into the picture. It is widely held and strongly maintained by FANS 1 supporters that necessary modifications and costs to upgrade the ACARS-based systems to make them fully compatible with ATN when ATN becomes available, will be within reason. Others argue however, that these modifications may not be so easy or inexpensive and that there are other interim steps that could be taken while ATN is under development. They further fear that using the ACARS as an interim system could hurt the prospects of full worldwide ATN implementation because it would encourage a piecemeal approach around the globe, thereby reducing the foreseen benefits of the fully integrated ATN. Those who argue for a go-slow approach to implementation of the ACARS-based systems also raise issues of integrity, robustness and other safety related issues.

For those promoting the early implementation of CNS/ATM based on ACARS, it is believed that its use will permit an expeditious and orderly transition to the ATN bit-oriented standards-based networking environment and allow an orderly evolution. Furthermore, they claim there are thousands of aircraft currently equipped with the ACARS data link, which is not unusual given the need for reliable AOC and AAC.

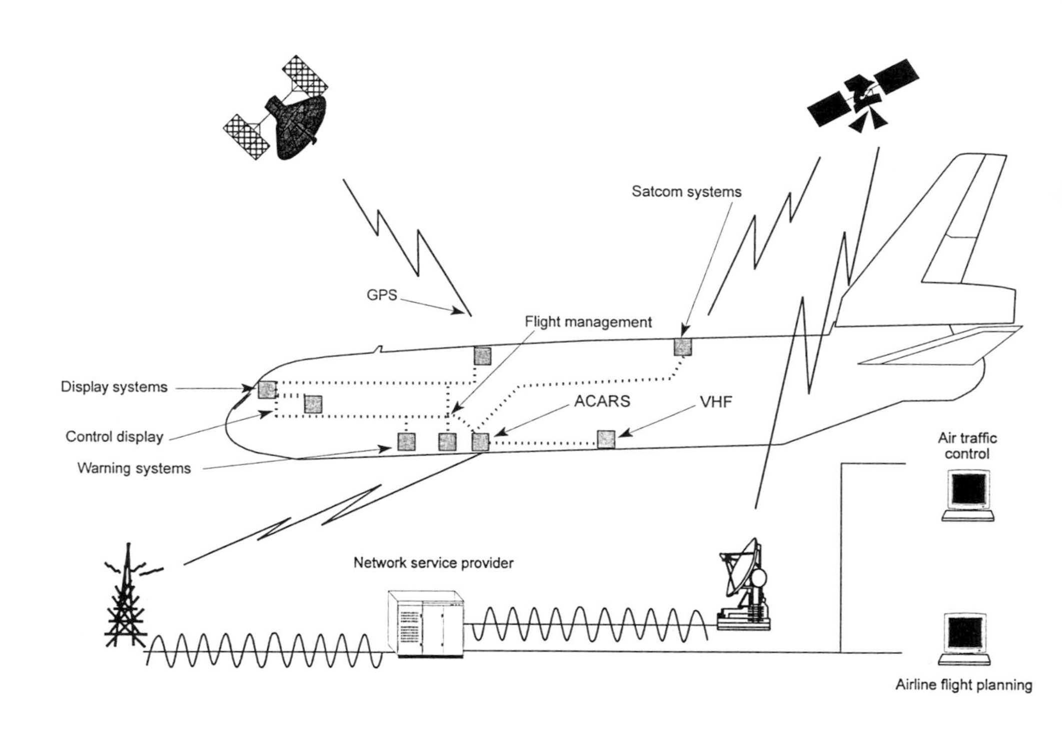

Figure 3.6 FANS 1 functionality

The Fourth Meeting of the ICAO FANS II Committee did, however, endorse the prompt use of the ACARS-based systems as a migration path to the ATN, believing that the corresponding benefits could be achieved incrementally. It was further agreed that when a bit-oriented air-ground data link, which can convey both code and byte independent information becomes available, then the ACARS-based system would not be necessary.

This view has been tentatively supported by a few ICAO regional planning groups and fully endorsed and planned for implementation by the Asia/Pacific Air Navigation Planning and Implementation Regional Group (APANPIRG).

The Air Navigation Commission, when reviewing the work of the Aeronautical Telecommunications Panel (ATNP) agreed that early use of ACARS-based systems could serve as a valuable step toward the early introduction of ATM applications. In follow-up, ICAO circulated a letter to all States on policy issues concerning planning and implementation of CNS/ATM systems (see Appendices E and F).

Conclusion (FANS 1) Packages similar to the FANS 1 for the 747-400 are being developed by Boeing for application on a range of other Boeing aircraft including the 737, 757, 767 & 777.

United Airlines is one of four Asia-Pacific carriers that prompted Boeing and avionics manufacturer, Honeywell, to develop the software package for the 747-400 flight management system (FMS) that would allow early advantage to be taken of CNS/ATM systems. The initial aims in modifying the FMS were to reduce the number of errors generated by navigation deviation or blunders, to eliminate transmission and interpretation errors by using data link and to introduce an initial form of oceanic automation by using ADS. The goal was to reduce costs by using existing hardware.

The FANS 1 package combines ATC data link communications, global navigation satellite systems (GNSS), primary navigation and ADS capability in a single box. Other features include direct controller/pilot data link communication (CPDLC) and flight plan uplink with auto-load (Aleshire, 1994b). These features mean that FANS 1 equipped aircraft will ultimately be capable of dynamic random routing operations in airspace without fixed airways or terrestrial navaids, such as over the oceans. Integration of ATC data link into the FMS means that ADS messages do not have to be laboriously assembled from several different sources. For example, the primary source of position and time data: the GPS (GPS is a GNSS provided by the United States and is described in more detail in Chapter 4), is itself integrated into the FMS. The FMS receives and displays ATS messages on a multi-function control/display unit while all ATC data link messages are archived by the FMS and individual messages could be selected and printed by the flight crew. Figure 3.7 displays the data link interface on the Boeing 777 as part of the FANS 1 package.

The Future Air Navigation System

Airbus' AIM FANS A and B

Airbus Industry has launched a programme providing airlines with a flexible approach to equipping aircraft for navigating in a satellite-based airspace environment. The Airbus Interoperable Modular (AIM-FANS), as it is called, is designed to offer new communications, navigation and surveillance functions on all Airbus Industry fleets (A300/A310, A319/A320/A321 and A330/A340). In the view of Airbus, each area of airspace has specific characteristics, thus the implementation programme will vary from one region to another and that manufacturers must therefore adopt a flexible approach. Additionally, Airbus believes that each step, in each region, must offer some benefits to the user in terms of reduced flying time, with the associated extra payload or fuel savings, in order to justify the investment (Signargout, 1995).

To accommodate Airbus objectives, AIM-FANS systems will be upgraded with future developments in mind. This will ensure that evolutionary changes can be incorporated in a straightforward manner, without the need for a complete system redesign at different implementation phases. The target is to have the first system upgrade, known as the FANS A, available for the A330/A340 by mid-1997 to satisfy the needs of the long-range Airbus operators in the Pacific basin.

The first level of data link function for ATC purposes, based on the Arinc Specification 622 protocol (ACARS), is planned to be introduced by the end of 1996. This upgrade will include the development of FMS/ATC functions, the development of the human-machine interface for data link and introduction of the new airborne ATS data unit (ATSDU). Evolution towards full implementation of the communications system will involve the introduction of ATN protocols and the upgrade of the communication media.

The Airbus FANS A will therefore be offered on a step by step modular basis. The programme is divided into two parts: FANS A and B. The primary A package uses the basic CNS elements already developed. The later FANS B package will be based around a more mature set of CNS systems. The timing of the second generation of FANS availability will depend on ATN capability.

- MFD display
- Cursor control selection
- Functionally similar to 747-400

ATC	FLIGHT INFORMATION	COMPANY
REVIEW	MANAGER	NEW MESSAGES
	ATC MENU	
ALTITUDE REQUEST	WHEN CAN WE EXPECT	EMERGENCY REPORT
ROUTE REQUEST	VOICE CONTACT REQUEST	ALL REQUESTED REPORTS
SPEED REQUEST	LOGON STATUS	POSITION REPORT
CLEARANCE REQUEST		FREE TEXT MESSAGE
		EXIT MENUS

- Route request via MFD form
- Most pilot selections via form

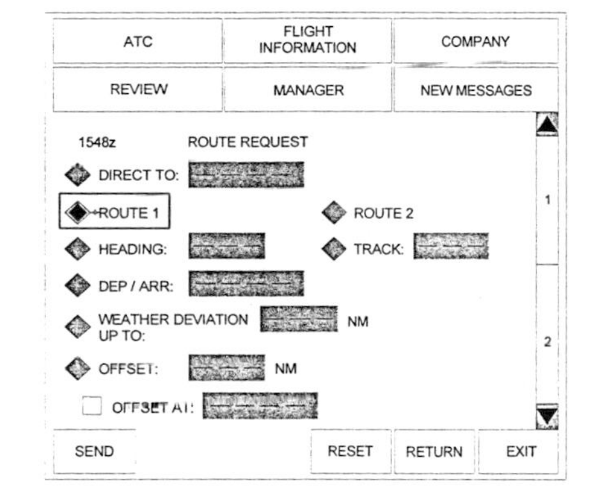

- Full display of uplink
- Color coded display
- Action selection via CCD

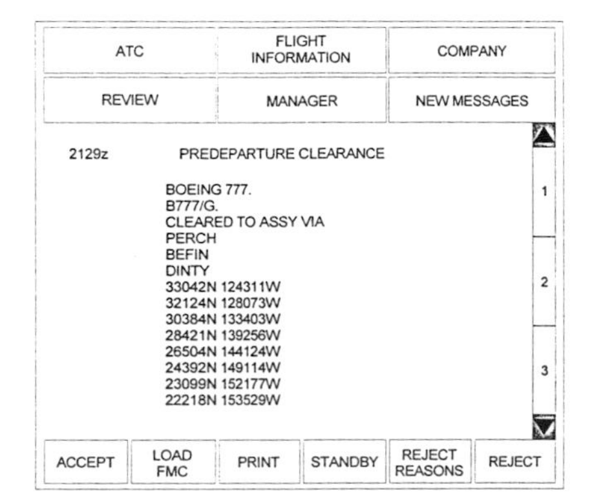

Figure 3.7 Data link interface on the Boeing 777

McDonnell Douglas' FANS 1/A

McDonnell Douglas began its future air navigation efforts with a programme to develop a FANS-compliant DC-10 test platform with United Airlines. The primary features of the upgrade, like its two competitors, will include GPS navigation and high gain satellite communications data. Information management and a graphic real-time weather display may be included in longer term plans for the MD-11. The newer package for the MD-11 will be known as the FANS-1/A and was expected to be available in 1996 (Norris, 1994a).

Regional airlines and FANS capabilities

A GLOBALink/CNS service was unveiled in July 1994. It is known as the CNS-10 avionics unit and is produced by Aeronautical Radio (ARINC) and Magellan Systems. The CNS-10 incorporates a VHF radio, an aircraft communications, addressing and reporting service (ACARS) unit, a control and display unit and a GPS receiver all in a 3.7 kilogram box (Daly, 1994b). The first uncertificated CNS-10 was delivered in Spring 1995 with a certificated version by the end of 1995. The expected market in the United States alone is 100,000 units for general aviation aircraft. CNS-10 is expected to bring the benefits of CNS/ATM to smaller aircraft at an affordable price.

ATC ground infrastructures

The success of implementation of CNS/ATM will be based on the introduction of enhancements in a coordinated way both in the air and on the ground. Upgrades in the air will not take place unless the ATC systems on the ground are adequately upgraded to provide the advanced services that CNS/ATM is capable of offering. Both sides will therefore want to know that the other is committed to making the expenditures necessary for success of the future system.

Before investment and implementation can begin, a clearly defined plan must be developed. This is where the ICAO regional planning and implementation groups described in Chapter 2 come in. Through their participation in regional planning activities conducted through the established regional planning processes, States are planning for implementation of CNS/ATM systems. The regional planning groups are now in the process of updating or developing new comprehensive regional air navigation plans (ANPs) (ANPs are described in Chapter 2), that not only describe the basic operational requirements and planning criteria of CNS/ATM systems, but also list the implementation requirements for the facilities and services that individual States will be responsible for, in accordance with their obligations under Articles One and Twenty Eight of the ICAO Convention.

The evolution of North Atlantic oceanic data link capabilities On long oceanic flights, the controller and pilot rarely communicate directly with each other, but usually communicate via HF radio indirectly through HF operators who relay messages. On oceanic crossings, a clearance is issued prior to entering oceanic airspace. In 1985, Canada began delivering these oceanic clearances to North Atlantic flights, for the first time, via VHF data link communications. The controller had to manually enter the clearance into a computer terminal, after which a data link system delivered the clearance to the flight. Within a year, the input of the clearance had been automated and the coverage of the data link had been significantly expanded. By mid 1995, approximately 36 percent of the flights entering the oceanic airspace controlled by the ATC Centre at Gander obtained their clearances in this manner (Walker, 1993).

In June, 1995, Transport Canada agreed to a pre-operational trial with one airline , eventually expanding to several airlines, to determine the viability of using HF data link to deliver waypoint position reports over the North Atlantic Ocean to Gander Oceanic Area Control Centre (OACC). So far the trial has proven to be very successful. It is expected that all HF voice position reports will be made in this way by aircraft while in Gander OACC.

The Gander automated air traffic system will process the position report messages and automatically update flight progress data. When the aircraft passes a point stored in the flight management computer (FMC), it automatically initiates the position report. Position report data is not entered manually, but is automatically inserted in the message from the FMC, so there is a safety benefit in that waypoint insertion errors can be reliably detected. There will eventually be a benefit in frequency spectrum efficiency and in radio operator workload.

Iceland began experimenting with delivering oceanic clearances to North Atlantic flights operating in their area of responsibility, by VHF data link in July 1991, which also included experimentation using satellite data link for delivery of these clearances.

In addition to the delivery of clearances, several States in the North Atlantic Region are experimenting with ADS, where information from aircraft avionics systems is passed to ATC units using the capabilities provided by data link. This information is then used to depict the flight on a display, somewhat like it would be on a radar display, or the information is used for other purposes. Because ADS is considered as a sub-function of the surveillance component of CNS/ATM systems, it is discussed in greater detail in Chapter 5.

Experiments continue to be conducted by several North Atlantic ATC service provider States to test and verify changes in operational practices required by transition to a new CNS system. Planning for the requirements for implementation of CNS/ATM in the North Atlantic is accomplished through the North Atlantic Systems Planning Group (NAT SPG). The NAT SPG is one of the ICAO regional planning groups described in Chapter 2 and is composed of the States that provide ATS to flights over the North Atlantic Ocean.

Pacific Ocean milestones After many years of intense planning, commercial satellite data communication was inaugurated on 27 September 1990 on a United Airlines Boeing 747-400. By March of that year six 747-400s operated by Qantas Airways, Japan Airlines and United Airlines were using the service on an operational basis. Airlines' CNS/ATM activities in this region are currently based on the FANS 1 package described in the paragraphs above.

United Airlines, which has invested a great deal of time and money in its examination of data link applications, claims there are great savings to be had. United has successfully demonstrated the capability of ATC satellite communications on its trans-Pacific routes. Using data link to request ATC clearances, the carrier has been able to save fuel and money (Nickum, 1992). According to Nickum, "the United Airlines' pilots working with the system have been very enthusiastic about its capability and have found that a clearance request usually requires no more than a minute instead of as much as twenty minutes when using HF voice communication". Furthermore, it is claimed that the system is user friendly and provides data quickly to the pilot, resulting in better situational awareness, especially with regard to weather and flight planning. Using the system, the pilot can often request clearances and arrive at decisions based upon desired aircraft performance, which has a direct impact upon fuel consumption. United claims to have achieved typical fuel savings of 1,500 to 1,800 kilograms per flight by optimizing route and cruise climb. Additionally, United expects to be able to halve the contingency fuel which its Boeing 747-400s carry from Los Angeles to Sydney.

Nickum states that the whole approach to FANS 1 was accomplished by a team of players, coordinated by the Informal South Pacific Air Traffic Control Coordination Group (ISPACG). Members of ISPACG include the aviation administrations and agencies of the States responsible for the provision of air traffic services in the area of the South Pacific (i.e., Australia, Fiji, French Polynesia, New Zealand, Papua New Guinea and the United States). These States began making the necessary changes to their ATC systems that would be required for CNS/ATM implementation which subsequently led to the development of the FANS 1 package. Other major players included Honeywell, who worked with Boeing in developing modifications to the flight management computers, and the four major airlines who participated in the development and certification process.

The interim FANS implementation adopted by ISPACG is designed to evolve gracefully into the final FANS CNS/ATM implementation, after that has been designed and requirements have been validated. The objective is to have a minimum of throw away design.

The Russian Federation The vast airspace over Siberia and parts of the North Pole have been alluring to the airlines for years. In previous times, this expansive airspace was prohibited for use because of military restrictions imposed by the former Soviet Union. As political realities have evolved, the major representative of international airlines, the International Air Transport Association (IATA), has made the use of this airspace a strategic objective (IATA, 1995b). However, except for the limited use of a few routes, and even that with a very low capacity, no progress in implementing a route network has been achieved because of the almost total lack of a suitable aeronautical communications infrastructure to support international operations.

The Russian Federation civil aviation authorities have therefore embarked on a federal modernization programme that will provide for the phased transition to a future CNS/ATM based system by the year 2005. The programme calls for a phased transition from the existing aeronautical systems to an integrated CNS/ATM system based on the ICAO concept. CNS/ATM system priority development will be toward satellite communication systems in addition to VHF data link. The Russian

Federation expects to realize a five to one cost/benefit ratio through the early implementation of CNS/ATM technology (ICAO, 1995d).

A number of CNS/ATM specific routes over the Russian Far East linking North America and the Orient have already been developed and flight tested. Eventually, these new routes and the implementation of CNS/ATM technology will allow non-stop service, for example, between the East Coast of the United States and China, or between the Mid Western United States and Hong Kong. Much of this work is based on related progress in the Pacific Region which is itself based on the FANS 1 package described earlier. Because of the progress made by the Russian Federation by the beginning of 1996, one major airline, British Airways, has decided to equip its entire fleet of Boeing 747-400s with the FANS 1 package.

Conclusions (glance at the future).

Today, several States responsible for providing ATS in the Pacific and in the North Atlantic Regions, are combining their efforts in order to obtain early benefits from CNS/ATM systems. Initial benefits from CNS/ATM implementation are expected to be had in the airspace over the North Atlantic and Pacific Oceans. The vast airspaces over Siberia and the North Pole, previously unusable for international flight, are being developed based on CNS/ATM technology. One of the early benefits of these efforts is that data link capabilities to communicate with aircraft are being put to increasingly greater use.

The eventual widespread adoption of satellite communication technology is expected to provide a new measure of safety in oceanic regions, to the benefit of ATS provider States and the entire aviation industry (Gribbin, 1991).

Finally

* Orders for the FANS 1 upgrade for the Boeing 747-400 flight management system have exceeded 200 and increasing by the end of March 1995. The upgrade for the Boeing 777 was expected to be certificated by the fourth quarter of 1996. Boeing is also studying versions for the 737, 757 and 767 aircraft. It is estimated that by the end of 1996, seventy five percent of aircraft operating in the Asia Pacific Region will be FANS 1 equipped. By the end of 1998, more than 600 aircraft are expected to be equipped;

* By April 1995, a growing number of States already had committed to installing ground systems to support the FANS 1 package. By September of 1995, New Zealand had the world's first operational CNS/ATM ground system in place, developed by CAE Electronics. The other States scheduled to install these systems are the member states of ISPACG (Australia, Fiji, French Polynesia, New Zealand and the United States). Additionally, the Russian Federation, India, Indonesia, Japan, Korea, Singapore and Thailand have made commitments to such systems;

* In mid 1996, ARINC offered to develop solutions that would give FANS capabilities to classic aircraft, including older B747s, DC10s, L101s, A300s and A310s;

* Considerable progress has been made with regard to ICAO Standards and Recommended Practices (SARPs) for CNS/ATM systems. SARPs for two data links or sub-networks, AMSS and SSR Mode S, have been written. Work on enhanced modes of VHF data link (VDL) to support communications as well as navigation requirements continues and SARPs are being developed with a target completion date of 1999;

* The Aeronautical Telecommunication Network Panel (ATNP) is developing the international SARPs, procedures and guidance material under which the ATN will be implemented. At its first meeting in June 1994 the ATNP endorsed two documents which cover exchange of ATS messages over the ATN (ICAO, 1995c). The first set of ATN SARPs will be considered at the second meeting of the ATNP scheduled for late 1996;

* Data link applications are operational in many locations and satellites for voice and data link communications are currently available. Their capacity is expected to expand by a factor of thirty by 2010 and aircraft equipage is expected to be at 8000 by the end of the next decade. Aircraft equipage for VHF data link is expected to be at 4700 aircraft by the mid 1990s and for Mode S it is anticipated to be at 3,000 aircraft by the year 2000 (ICAO, 1993a);

* Communications for ATS via VHF data link based on ACARS is being experimented with for specific purposes within portions of North America, Europe, Asia and the Pacific. Communications for ATS via Mode S data link will be initiated in North America in the mid 1990s, within the Pacific region in the late 1990s and within Europe in the early 2000s (ICAO, 1993a). The potential of HF data link to support ATM applications is being investigated by the AMCP and results are expected by the fourth meeting of the AMCP by the end of 1996;

* Operational use of the ATN is expected to be initiated in North America, Europe and the North Atlantic regions by the late 1990s and in the Pacific region in the early 2000s.

Questions and exercises to expand your knowledge

1) Describe a complete air-ground communication system making use of the elements envisaged with the implementation of CNS/ATM systems. Include the following in your answer:

   * ATN;
   * Digital Data Link;
   * HF radio band;
   * VHF radio band;
   * Satellites;
   * Mode S Data Link.

2) Draw a simple diagram of the system you described above.

3) What are the major differences between the communication systems that are in use today and the ones foreseen using CNS/ATM technologies?

4) What are the main advantages of the communication systems envisaged using CNS/ATM technologies?

5) Identify the types of information that could be passed from airborne to ground based systems automatically. How could this information be used by an advanced ATM system toward the provision of a more efficient and effective service?

6) This chapter lists three main shortcomings in present systems and describes them as "inherent" in the systems themselves. What is meant by inherent?

7) The FANS Committee recognized that the communications shortcomings identified would not be the same for every area of the globe. Give an example of this.

8) Why does terrestrial communication remain necessary in the future system?

9) Research the International Standards Organization's (ISO) Open Systems Architecture (OSI) and explain the concept in greater detail than it is explained in this chapter. Why is the OSI so important in the CNS/ATM environment?

Canadian airspace management simulator providing full-scale simulation for training controllers and development of operational procedures. (Picture provided courtesy of CAE Electronics Ltd.)

4. Navigation

Introduction

> Probably right from the time man...started to wander around the earth he's been looking for some simple way to figure out where he was and where he was going....It's such a basic problem you'd think we'd have come up with something that really works (Hurn, 1989).

Locating our position on Earth and determining the course to steer to arrive at the next desired point has always been an obscure science, but one that has fascinated man from earliest times. According to Hurn, it probably began with early travellers marking their trails with piles of stones or some similar method, with limited success. As man began exploring the oceans, celestial navigation was all that he could rely on, and even this was an inexact way of determining his position.

In modern times, several methods of navigation were developed, making use of electronic means and instruments with great success around coastal waters and over land. But an exact system for navigating reliably over all parts of the globe did not come into man's reach until very recently.

As explained in the previous chapter, the use and integration of communication, navigation and surveillance systems will evolve to be vastly different than it is today, due primarily to a much greater use of satellite technology. This is certainly true in the realm of navigation. In fact, newly developed methods for pinpointing one's exact location anywhere on the earth and then easily navigating to a new location can be seen as the culmination of several developments in two separate fields: satellite and computer technology. It is the merger of the two, along with the perceived cold war military requirements of both the United States and the former Soviet Union, that led to the development of two autonomous satellite navigation systems.

The system developed by the United States has been designated as the Global Positioning System (GPS), while the one developed by the former Soviet Union has been labelled as the Global Orbiting Navigation Satellite System (GLONASS). Full deployment of GPS has now been achieved at twenty four satellites. The GLONASS has also achieved full deployment of twenty four satellites by mid-1996. All of these satellites are high enough that they can avoid the problems encountered by land based systems. Both systems utilize technology accurate enough to give pinpoint positions, almost anywhere in the world, at any time.

ICAO has given the generic term: Global Navigation Satellite System (GNSS) to the concept of navigation by satellite. GNSS encompasses satellite constellations, aircraft receivers, and system integrity monitoring. It therefore includes the two systems that currently exist in addition to any satellite services that may be used to augment these systems.

In any discussion of GNSS, and certainly in the case of this text, direct reference must often made to the two existing systems described above as these are the first two truly navigation satellite systems on the scene. Furthermore, development and implementation of GPS is presently further along than is that of GLONASS, and much of the research presently being conducted and data collected refers to GPS, thereby necessitating frequent reference to this particular system.

The ICAO concept is quite definite in its preference of GNSS for navigation, seeing it as providing independent on-board position determination, and being a key feature of future CNS/ATM systems. GNSS is expected to evolve to become a *sole means* of navigation, and eventually replace the presently used long-range and short-range navigation systems. It is expected to provide global coverage and to be accurate enough to support enroute navigation and meet non-precision type approach needs and eventually the full range of precision approach requirements in the future (ICAO, 1993a). Institutional and technical obstacles are presently being addressed by the international community under the auspices of ICAO, aimed at quickly bringing about the anticipated benefits of GNSS to the international civil aviation community.

This chapter discusses GNSS, how it works, its potential uses and some of the issues that lie between the present status of the two existing systems and their implementation as full civil aviation global satellite navigation systems. Additionally, background information is provided on the work of the FANS Committee, ICAO and its Panels which includes the identification of functional shortcomings of present navigation systems, the envisioned changes that will form the basis of future navigation systems and methods, and the benefits that would come to air traffic management (ATM) as a result of this new system of navigation.

The concept of Required Navigation Performance (RNP), as developed by the FANS Committee, which will define the equipment requirements for performance of aircraft in a particular airspace or phase of flight, is examined in some detail as it will have a significant impact on equipment carriage in aircraft, airspace planning and development of infrastructure.

This review will also discuss some of the ways and places where the navigation element of CNS/ATM is being experimented with and implemented.

Definitions

In order to fully understand this chapter, it is necessary to first become familiar with a few of the common international terms which are frequently used when referring to navigation and future systems and which you will see often as you become involved with international civil aviation. The following definitions are taken from various ICAO manuals, some mature and others still under development. In fact some were developed specifically for the FANS Committee to do their work and may be further refined by ICAO (ICAO, 1993a).

Global Navigation Satellite System (GNSS) A worldwide system of position and time determination, that includes one or more satellite constellations, aircraft receivers, and system integrity monitoring, augmented as necessary to support the required navigation performance for the actual phase of operation.

Area Navigation (RNAV) A method of navigation which permits aircraft operation on any desired flight path.

Differential GNSS (DGNSS) DGNSS is an augmentation, the purpose of which is to determine position errors at one or more known locations and subsequently to transmit information derived to other GNSS receivers to enhance the accuracy of the position estimate.

Wide Area differential GNSS (WADGNSS) WADGNSS is a differential (sometimes referred to as augmented), GNSS where the differential corrections are usable over an extensive geographical area for the supported phases of operation.

Local Area differential GNSS (LADGNSS) LADGNSS is a differential or, augmented, GNSS in which the differential corrections are usable for the supported phases of operation within a limited geographical area.

Primary-means navigation system A navigation system that, for a given operation or phase of flight, must meet accuracy and integrity requirements, but need not meet full availability and continuity of service requirements. Safety is achieved by either limiting flights to specific time periods or through appropriate procedural restrictions and operational requirements.

Receiver autonomous integrity monitoring (RAIM) RAIM is a technique whereby an airborne GNSS receiver/processor autonomously monitors the integrity of the navigation signals from GNSS satellites.

Required Navigation Performance (RNP) A statement of the navigation performance accuracy necessary for operation within a defined airspace.

Sole-means navigation system A navigation system that, for a given phase of flight, must allow the aircraft to meet all four navigation system performance requirements - accuracy, integrity, availability and continuity of service.

Supplemental-means navigation system A navigation system that must be used in conjunction with a sole means navigation system.

Sole means of navigation as described above, apply only to aircraft avionics equipage and not to ground infrastructure that support other navigation means such as VOR/DME. Thus it intends to imply that an aircraft can legally fly with only one type of avionics although it may carry others as well.

The Review of the General Concept of Separation Panel (RGCSP) is expected to review the definitions of RNP to include the full spectrum of required characteristics which would take into account integrity and availability.

Shortcomings of the present system

> The present systems, introduced in the 1940s...require thousands of air traffic control (ATC) units, ground-based VHF relay stations and an extensive network of navigational radio beacons strung around the globe.... At the same time, the system is unable to grow to meet increased congestion around airports...." (Campbell and Salewicz, 1995, p. 21).

The previous chapter described the shortcomings of present communications, navigation and surveillance systems as identified by the FANS Committee, which recognized that for an ideal worldwide air navigation system, the ultimate objective would be to provide a cost-effective and efficient system that would permit the flexible employment of all types of operations in as near four dimensional freedom as their capability would permit.

FANS I (ICAO, 1988b) described the shortcomings of communications, navigation and surveillance systems around the world as amounting to essentially three factors as follows:

* The propagation limitations of current line of sight systems;

* The difficulty, caused by a variety of reasons, to implement present CNS systems and operate them in a consistent manner in large parts of the world;

* The limitations of voice communication and the lack of digital air to ground data interchange systems to support automated systems in the air and on the ground.

Navigation

Ideally, aircraft want to fly the most fuel efficient routes between their points of departure and their destination. For a variety of reasons which include technical capabilities of aircraft, environmental concerns, national security concerns and the inability of the ATC system to safely separate and otherwise handle large amounts of traffic on random routings, a series of ATS routes or airways are established around the globe. In fully developed ATM systems, traffic on direct routings will be more easily accommodated. Chapter 6 discusses ATM issues in greater detail. For the purposes of this chapter, it is important to know that suitable navigational guidance, defining the centre lines of each of the established routes or airways, must be available (ICAO, 1984).

The most commonly used means of navigational guidance for areas over land and for short routes is the ground based, point source navigation aid (ICAO, 1984). The one most commonly used is the very high frequency (VHF) omni-directional radio range (VOR) supplemented by distance measuring equipment (DME). In many other cases, however, non-directional radio beacons (NDB) are used.

Navigational guidance over the high seas is generally provided from two different sources:

* Ground based long-range navigation systems providing an area type coverage. The two most commonly used systems are known as LORAN-C and OMEGA; and,

* Self contained navigation aids which are practically independent from externally derived navigation inputs. These are more commonly known as inertial navigation systems (INS).

Area navigation (RNAV)

The method used for navigation over the high seas, described above, is commonly known as area navigation (RNAV). RNAV has been available and used quite extensively for many years not only over the high seas but also over continental airspace. In fact, modern aircraft are increasingly equipped to utilize newer techniques based on RNAV.

In general terms, RNAV equipment operates by automatically determining aircraft position from one or more of a variety of inputs such as VOR, DME, OMEGA, LORAN-C, INS, and most recently, from satellites.

Distances along and across track are computed to provide the estimated time to a selected point together with a continuous indication of steering guidance. RNAV allows flight to be conducted along any track, circumventing the need to fly directly over ground-based navigation facilities such as the VOR or NDB. Therefore, RNAV equipped aircraft do not need airways or routes, however, when an airway or route is established specifically for RNAV equipped aircraft, it is typically designed to be as near as possible to the most direct link between origin and destination (ICAO, 1991c).

Area navigation or RNAV based on satellite technology is a key feature of CNS/ATM from which many of the other benefits will flow. To take advantage of the opportunities, aircraft will increasingly be RNAV equipped. In the near term, RNAV will be supported by a variety of navigation inputs (ICAO, 1991c). It is therefore necessary to briefly discuss the general characteristics of RNAV airborne equipment.

VOR/DME The least complex of VOR/DME type equipment for RNAV use are those using station moving. In effect, this type of RNAV electronically offsets a selected VOR/DME facility (by a range and bearing calculated by the operator) to the position of the next waypoint and the aircraft is then provided with apparent VOR steering guidance to that waypoint.

The disadvantages are that the equipment is subject to the operational coverage and reception limitations of the selected facilities and any other errors inherent in the system.

OMEGA/very low frequency (VLF) OMEGA is a global navigation system which operates in the hyperbolic mode. The accuracy depends upon the geometry and the quality of the reception of OMEGA and/or VLF signals from the various stations. There are eight of each type of OMEGA and VLF globally dispersed transmitting stations. Airborne systems may use OMEGA alone, or a combination of the two types of signals for the provisions of position and tracking guidance. OMEGA can be a stand alone system providing the pilot with data, but in most modern aircraft the systems are duplicated, integrated with other systems on-board the aircraft and coupled to the autopilot.

The disadvantages lie in the propagation anomalies and the spread of navigation accuracy resulting from such anomalies. OMEGA must therefore be cross-checked from time to time with conventional aids.

LORAN-C LORAN-C is a radio navigation system which uses time-synchronized time signals from ground transmitting stations spaced several hundred miles apart. Aircraft position is determined by measuring the difference in arrival time of LORAN-C pulses from three or more ground stations. As with OMEGA, LORAN-C equipment may be stand-alone, but modern systems are more frequently integrated with a navigation computer in order to provide a range or positional and associated information, and coupled to the autopilot.

The disadvantages of LORAN-C stem from the fact that it is subject to local interference; a failure of one transmitter can leave a major area without coverage; and, approval of LORAN-C for RNAV operations must be limited to the geographical area of good ground wave signal reception.

Inertial navigation systems (INS) The INS is a totally self-contained equipment that operates by sensing aircraft accelerations with a gyro-stabilized platform. Output functions of the system include accurate present position information, navigation data, steering commands and angular pitch, roll and heading information. Most aircraft fitted with INS have a duplicated or triplicated system. The normal operating practice is to input the system with the aircraft known position with a high degree of accuracy prior to departure from the aircraft stand. By presetting a series of waypoints, the system will navigate the aircraft along a predetermined track. Waypoints are usually fed into the system prior to departure, but new points can be inserted at any time.

The major disadvantage of INS is that accuracy becomes degraded with elapsed time since the last update for which a linear decay of 1.5 to 2 nautical miles per hour must be allowed. Additionally, INS can be expensive, especially when considering sophisticated updating capabilities and necessary double and triple redundancy.

DME/DME The most accurate means currently available of updating RNAV and flight management system (FMS) equipment within continental airspace is by reference to multiple distance measuring equipment (DME) stations on the ground, with a minimum of two suitably positioned facilities being needed to provide a position fix.

The disadvantages, like those of the VOR, are associated with operational coverage and reception limitations.

Global navigation satellite systems (GNSS) One of the advantages of satellite navigation systems is that they allow independent navigation, where the user performs on-board position determination from information received from transmissions by a number of satellites. Similar to other types of RNAV, GNSS equipment on-board the aircraft can be stand-alone, providing the pilot with position data, from which navigation can be carried out, or it can be integrated with and/or supplemented by other equipment, such as the FMS, from which navigation and steering guidance can be provided or, for instance, by INS, whereby GPS would be used to update the INS.

Navigation using GNSS is not very different from the other methods of navigation described above. The sensor inputs are provided by satellites and these allow present position to be determined. The pilot, or on-board computer systems use the inputs to determine steering guidance to arrive at the next point.

Although GNSS is capable of offering highly accurate position determination, there are many factors that can introduce errors to the estimate of the user position. There are also institutional issues related to the fact that they are controlled, primarily, by military organizations, and even then, by only two States. These issues are being addressed by the international community through ICAO.

Navigation as envisaged in an integrated global CNS/ATM environment

The FANS Committee developed and expanded upon the roles of the various existing and unfolding navigation concepts and elements in a CNS/ATM environment (ICAO, 1988b) according to the groupings: *present radio navigation systems; satellite technology; approach and landing guidance systems; vertical navigation; and, required navigation performance capability (RNPC).* The foreseen roles and evolution of these elements are described below.

Present radio navigation systems

The present radio navigation systems serving enroute navigation and non-precision approaches (e.g., VOR, NDB) will be able to meet the RNP (RNP is explained in greater detail later in the chapter) conditions and co-exist with satellite navigation systems according to the FANS Committee (ICAO, 1988b). However, it is foreseen that satellite systems will eventually become the sole means of radio navigation and the present radio navigation systems will be progressively withdrawn. The timing of such withdrawal will depend upon many factors among which the implementation and quality of the new systems will be prominent, and will probably differ in various regions of the world. Because of its impact on international civil aviation on a worldwide scale, the transition and/or withdrawal will most likely be planned at the regional level through ICAO.

Satellite technology

The FANS Committee envisaged that developments in the use of satellite technology for aircraft navigation were such that satellite navigation systems would continue to evolve. Two basic concepts for satellite navigation were foreseen:

a) independent navigation, where the user performs on-board position determination from information received from broadcast transmissions by a number of satellites; and,

b) dependent navigation, using systems that provide radio-determination satellite services in which position, determined on the ground by multiple ranging measurements, are transmitted to the aircraft.

Systems of the first category will potentially provide highly reliable, highly accurate and high integrity global coverage independently and will meet the requirements for sole means of navigation for civil aviation. Although the RNP concept (described below) allows for more than one satellite navigation system to be in use simultaneously, from an aircraft equipment point of view, maximum interoperability is essential as it would significantly simplify avionics and thereby reduce cost. It would also be attractive if satellite systems could serve as a complement to and/or in backup role for each other.

Radio determination satellite services, mentioned in b) above lack frequency spectrum protection for safety services and they are also saturable, variable in their accuracy, and dependent on several communication paths. They are therefore, not considered adequate to meet global navigation requirements. GNSS as mentioned above, and as used in this text, will therefore refer to systems of the category described in a) above in which global coverage is provided.

Approach and landing guidance systems

In accordance with the needs of the international community, due to the fact that the instrument landing system (ILS) was outliving its useful life, or would soon do so, an ILS/microwave landing system (MLS) transition plan was developed which established a framework for coordinated worldwide transition from ILS to MLS, and adopted by the ICAO Council in 1987. Under the plan, it was envisaged that ILS would be largely replaced by MLS for precision approach and landing operations. When the FANS Phase I Committee completed its work in 1988, the ILS/MLS transition plan was still considered valid, while it was foreseen that GNSS would eventually provide adequate information to support non-precision type approaches and would also support the approach to the final approach fix for precision approaches.

In the ensuing years, several technological advances concerning the potential uses of GNSS as a precision approach aid had been realized. Concerns were therefore raised about the economic viability of MLS as a global replacement for ILS and, particularly, in light of the emergence of new technology systems with realistic prospects of cost-effective alternatives to ILS/MLS transition including GNSS.

In December of 1992, the ICAO Assembly addressed the aspects of ILS/MLS transition and called for the convening of a special meeting at the global level in order to assess the results of studies carried out (Kotaite, 1995).

As it became well accepted that GNSS could support precision approaches, the commitment of States, and in particular the United States, toward making the expenditures necessary for implementation of MLS systems, had eroded.

Based on the doubts surrounding the ILS/MLS transition plan and the Assembly's decision to have the matter reviewed, the ICAO Council convened a worldwide meeting of States in order to develop a new global strategy for transition to new precision approach systems that would take into consideration GNSS. That meeting, known as the Communications/Operational Divisional Meeting, held in March-April 1995, developed a new global strategy to guide States as well as industry for the next twenty years.

The global strategy agreed to (ICAO, 1995g):

* Retains the ILS as an international standard for precision approach and landing guidance for the foreseeable future;

* Encourages only limited implementation of MLS for precision approach and landing guidance for those locations where it is operationally required and economically beneficial;

* Promotes the continuing development of a multimode receiver or equivalent capability to maintain flexibility for aircraft operators in regions where there may be a mixture of ILS, MLS and GNSS approach and landing procedures; and,

* Promotes the ongoing research and development and validation for GNSS to be used for all weather approach, landing and departure operations.

As can be seen from the results, GNSS will play a significant role in worldwide development and implementation of precision approach systems. The ICAO All Weather Operations Panel (AWOP) (see Chapter 2 for a general description of ICAO panels) has been given the task of continuing to study this matter.

Vertical navigation

An ICAO Panel known as the Review of the General Concept of Separation Panel (RGCSP) (see Chapter 2 for a general description of ICAO panels) has been studying the relationship between barometric altimetry and vertical separation of aircraft above 29,000 ft. with the objective of determining the feasibility of reducing the presently applied minimum vertical separation criteria of 2,000ft. between aircraft. Below 29,000 feet, 1,000 feet vertical separation is applied. The difference lies in the fact that barometric altimetry is less accurate at higher levels. Barometric altimetry becomes even less functional at the very high altitudes at which future multi-mach aircraft will operate, in which cases geocentric altitude, based on GNSS measurement from satellite navigation systems is expected to be used.

Additionally, it is expected that geocentric altitudes could eventually serve as a cross-check on vertical position, especially between proximate aircraft.

Required navigation performance (RNP) capability

As explained earlier in this chapter, modern aircraft are increasingly equipped to utilize newer techniques, generally referred to as RNAV. As RNAV equipped aircraft are for the most part not dependent or constrained by the location of station referenced or ground based navigational aids, airways or routes serving RNAV equipped aircraft are typically designed to be as near as possible to the most direct link between origin and destination. It was recognized that this capability would facilitate a more flexible route system. The future navigation system will therefore be based on the availability of airborne RNAV capability.

A concept, primarily for use in airspace where adequate surveillance (surveillance is discussed in detail in Chapter 5) is available for ATC, known as RNP capability, was developed. RNP is broadly defined as the maximum deviation from assigned track within which an aircraft can be expected to remain within a given degree of probability. This concept avoids the need for ICAO selection between competing navigation systems, which had been the case in the past, however, it does not prevent ICAO from dealing with navigation techniques, especially those which are in wide use internationally.

The RNP concept was approved by the ICAO Council and assigned to the RGCSP for further elaboration.

Global navigation satellite systems (GNSS)

> Satellite navigation presents opportunities for standardized worldwide civil aviation operations using a common navigation receiver and for significant improvements in safety, capacity, service flexibility and operating costs. Adoption of satellite navigation systems will most likely lead to the eventual phase out of current ground based navigation systems (Dorfler, 1994, p. 1).

Aircraft avionics have evolved to such an extent that RNAV has been acknowledged by the FANS Committee as having the capability of offering improvements necessary to increase system capacity and efficiency and therefore, CNS/ATM systems will be based on the availability of airborne RNAV capability (ICAO, 1992b).

Campbell and Salewicz (1995) note that as early as 1966, the idea that satellites could be used to guide aircraft was gaining support. Advances initially took place in the military sectors of both the United States and the former Soviet Union with development of GPS and GLONASS respectively. GNSS will therefore initially be based on these two systems.

GNSS is expected to revolutionize the way that aircraft navigate, enabling them to operate in all types of airspace, in any part of the world, at any time, using satellite based avionics, eventually affording the possibility to States of dismantling at least a portion of their ground infrastructures (e.g., VOR, NDB, ILS).

The advantage to ATM will come from the fact that the accurate, reliable and continuously available position information available to the aircraft will also be made available to ATM ground systems through the advanced communication systems described in Chapter 3. Also, GNSS will offer the possibility to aircraft operating within an ATM system, within other constraints of that system, of adhering to any desired track using RNAV techniques based on GNSS input. In association with the RNP concept, these capabilities should permit a reduction in the separation standards used by ATC.

How it works

GNSS is based on satellite ranging. That means that position on earth is determined by measuring exactly how far something is from a series of satellites (Hurn, 1989). The system works by timing how long it takes a radio signal to reach the GNSS receiver on earth and then calculating a distance from that time. By ranging from three satellites a position determination can be made. Range measurements effectively establish the radii of spheres on which the user is located; the intersection of these range spheres establishes the user's position. For technical reasons, measurements from four satellites are needed to determine exact position, however, in theory, only three measurements are necessary. In fact, five will probably be needed for integrity reasons which are discussed later in this chapter. The basic principle behind GNSS therefore is that satellites are used as reference points for triangulating position on earth.

The GPS of the United States consists of twenty four operational satellites plus three working spares in six orbital planes, three globally dispersed monitor stations, three globally dispersed ground antennas and one master control station. A GPS user determines his position by processing range measurements to four satellites. Since the range is not a direct measurement of distance, but a measurement based on time, the term pseudo-range is used to describe the calculated range. By measuring pseudo ranges to four satellites and computing the satellite positions using ephemeris data transmitted by the satellites, the user equipment determines its three dimensional position and time. The user velocity is determined in a similar way by measuring range rates to four satellites. GPS provides both a standard positioning service (SPS) and a precise positioning service (PPS). The SPS is freely available to any civil, commercial or other user, internationally. The level of accuracy of the SPS is set by the United States military at 100 meters horizontal and 157 meters vertical at 95 percent probability (ICAO, 1993a), although in practice much greater accuracies are being realized. The PPS is a military position/navigation service providing accuracies higher than SPS through the use of cryptography (technical characteristics of the GPS are reproduced at Appendix G).

The GLONASS, initially developed by the former Soviet Union has now been taken over by the Russian Federation. GLONASS also consists of twenty four satellites with three in the standby mode. The user equipment operates in a passive mode and performs measurements on up to four satellites. A navigation message transmitted from each satellite consists of information on satellite ephemeric position and corrections relative to the GLONASS system time scale, as well as information concerning all satellites' condition. Based upon measurements, the user's three dimensional coordinates and speed vector components are determined and its time scale is referenced to that of the system (ICAO, 1993a).

Although there are some noticeable technical and operational differences such as the time a GLONASS satellite takes to complete a circular orbit, the inclination of the satellites and the number of orbital planes, the two systems essentially operate along the same fundamental principles (technical characteristics of the GLONASS are reproduced at Appendix G).

Weaknesses in GNSS systems

The two existing systems, despite being very powerful and offering several advantages over currently available ground based navigational systems, have limited ability to warn the users of malfunctions, which are generally referred to as problems of integrity. Overall, integrity refers to the trust which can be placed in the correctness of the information supplied by the total system. Additionally, the accuracy afforded to the user community is less than what is needed for the more stringent phases of flight associated with precision approaches. The limitations associated with accuracy will be overcome through differential techniques which are explained below, while the limitations associated with integrity will be overcome through one of two monitoring techniques referred to as receiver autonomous integrity monitoring (RAIM) and GNSS integrity channel (GIC) (ICAO, 1993a).

In addition to problems of integrity and accuracy, there are also questions related to continuity, availability (ICAO, 1994d) and the institutional commitment of the two provider States, which are currently being addressed. Continuity is the ability of the total system to perform its function without interruption during the intended operation. Availability refers to the ability of the total system to perform its function at the beginning of the intended operation (ICAO, 1994d). Availability can also be described as the ability of the total system to maintain the aircraft position within a total system error.

Continuity and availability obstacles will be overcome through augmentation methods also described below while institutional issues will require cooperation and agreement between States.

Integrity monitoring

There are two different methods for monitoring integrity of individual satellite ranging signals. One is based on observations of signals on the user side only; known as receiver autonomous integrity monitoring (RAIM). With RAIM, multiple position solutions are calculated and compared. A failed satellite signal can be detected by comparing the differences between these solutions.

The second method uses a network of ground stations monitoring the satellites' health. The ground monitoring stations transmit information to one or more master stations which derive the health status of the satellite signals and integrate information into a composite integrity signal. This signal is transmitted to the aircraft via a dedicated GNSS integrity channel. Both systems have advantages and limitations which are explained in more detail later in the chapter as part of the review of several of the programmes that are either in existence or are being developed to deal with integrity issues.

GNSS augmentation

The ICAO All Weather Operation Panel (AWOP), (ICAO, 1994d) describes specific augmentations as enhancements to the basic GNSS system that are intrinsically tied to satellite systems. GNSS specific augmentations include: differential techniques applied to the satellite signals, integrity determination, alerting systems and sources of additional ranging systems. Because of the widespread research, expected availability and potential of differential techniques for use with GNSS, it is important to have a basic understanding of this technology. Differential technology is briefly described below as defined by the AWOP.

Differential technology

The level of accuracy of GNSS varies for differing reasons, either because of technical problems associated with timing or because of institutional reasons as is the case with the SPS of the GPS provided for civil use by the United States military, which is a purposely degraded signal. To address these problems and make GNSS more accurate, differential technology is applied to the system. Differential techniques involve the transmission of a correction message, derived from measurements made on the ground, to users, via data link. By applying the correction information, the user can reduce the satellite ranging error (see Figure 4.1). The three areas being researched concerning differential techniques are as follows:

* Local area differential GNSS (LADGNSS) involves transmission of corrections to the code phase for each satellite. The corrections are derived from measurements collected at a ground reference site. As the ground reference sight is a known location, it can determine the inaccuracies of GNSS signals and compute differential correction messages, which would then be transmitted to the aircraft via a line-of-sight data link (see Figure 4.2);

* Kinematic local area differential GNSS (KLADGNSS) is similar to LADGNSS in the sense that corrections to aircraft position information are derived from a single-ground station. However the corrections are in the form of satellite signal carrier phase, thereby allowing the aircraft to employ interferometric principles for navigation;

* Wide area differential GNSS (WADGNSS) involves networks of data collection ground stations. Typically these ground stations might be separated by more than 500 miles or 1000km. Information collected at several ground stations is transmitted to and processed at a central facility to derive corrections to satellite range error component sources (e.g., satellite clock, satellite ephemeris and ionospheric delay). These corrections are then broadcast to users via a communications system which covers the data collection area (e.g., geostationary satellites or a network of ground transmitters) (ICAO, 1994d).

Additional ranging sources Geostationary and/or orbiting satellites, in addition to those of the GPS or GLONASS systems, can be used as additional ranging sources and can also serve to relay WADGNSS messages. This is discussed later in the chapter when reviewing specific concepts for the future. Figure 4.3 displays the GNSS architecture.

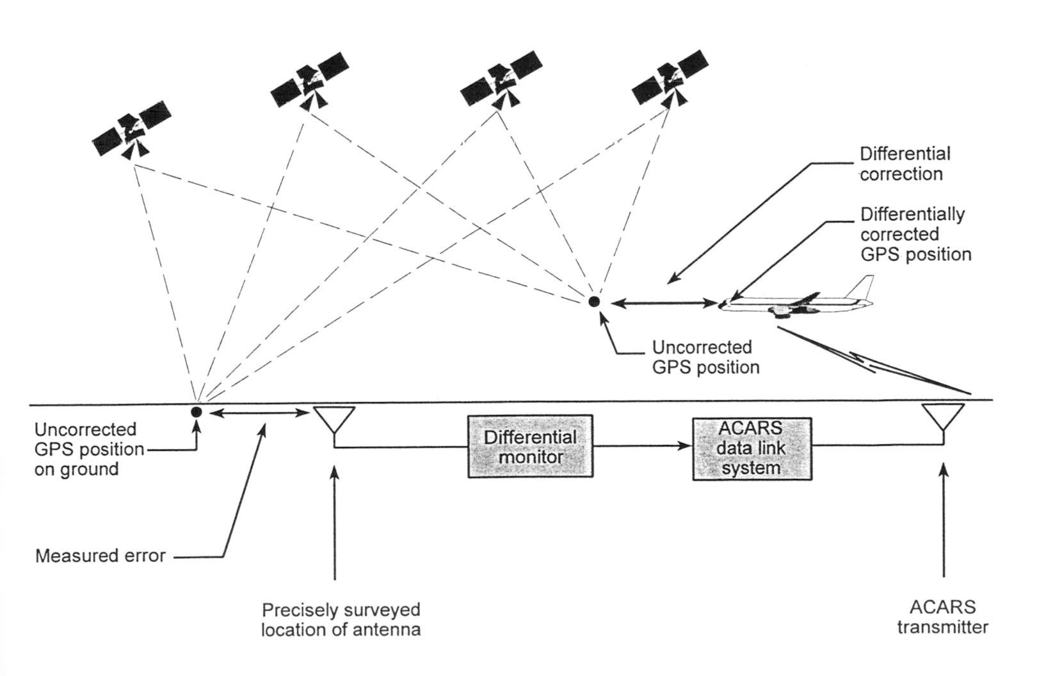

Figure 4.1 Differential correction

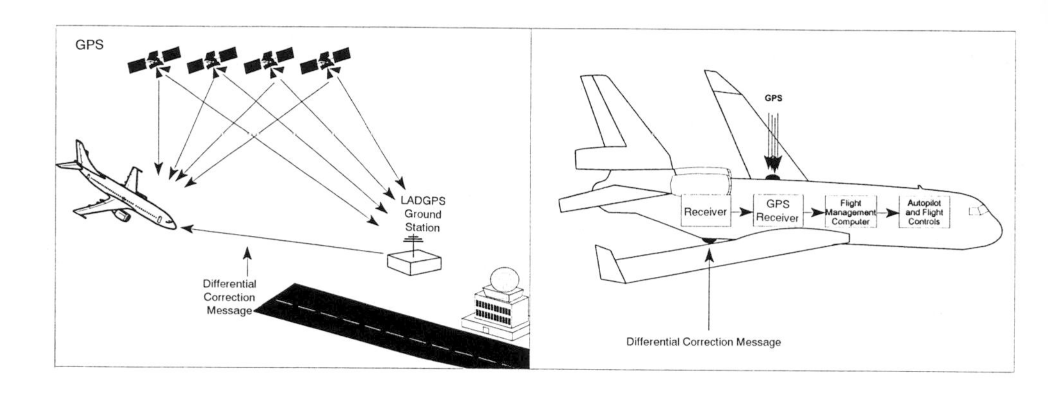

Figure 4.2 Local area differential correction

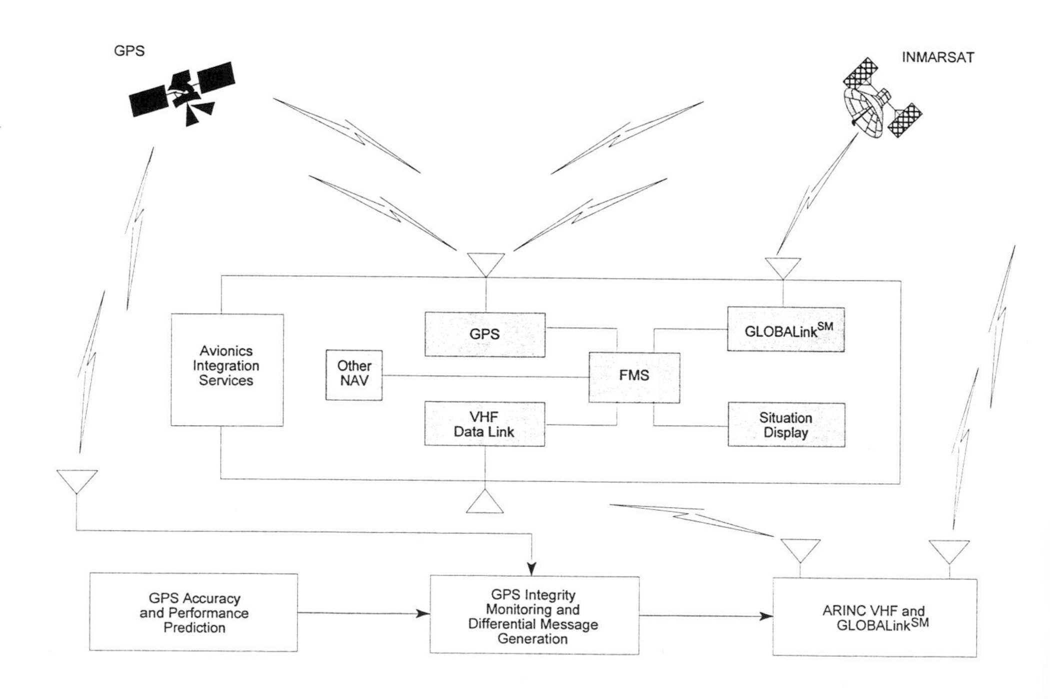

Figure 4.3 An example of GNSS architecture with integrity monitoring and differential correction

Required Navigation Performance (RNP)

GNSS will offer the capability of high integrity, reliable and accurate navigation. In order to maximize use of the technology for airspace planning purposes based on ATM requirements, the concept of RNP was developed by the FANS Committee.

The RNP concept was subsequently approved by the ICAO Council and assigned to the RGCSP for further development (see Chapter 2 for a description of panels). The RGCSP, among other things, developed the RNP manual (ICAO, 1994j) which describes the concept in detail. Excerpts of the RGCSPs work on RNP are reproduced below.

Introduction to RNP

The FANS Committee recognized that the method most commonly used over the years to indicate required navigation capability was to prescribe the mandatory carriage of certain equipment (ICAO, 1988b). This often constrained the optimum application of modern airborne equipment. Also, with satellites becoming available, the process would force a difficult selection process upon ICAO. To overcome these problems, the FANS Committee developed the concept of required navigation performance capability (RNPC).

The RGCSP, recognizing that capability and performance were different from each other, and that airspace planning is dependent on measured performance rather than on designed in capability, dropped the word "capability" and changed RNPC to required navigation performance (RNP).

The RGCSP developed the concept of RNP further by expanding it to be a statement of the navigation performance accuracy necessary for operation within a defined airspace. A specified type of RNP is therefore intended to define the navigation performance of the population of aircraft within a given airspace based on the navigation capability within that airspace. RNP type airspaces are identified by a single accuracy value. System use accuracy is based on the combination of the navigation sensor error, airborne receiver error, display error and flight technical error.

The RNP types specify the navigation performance accuracy of all the user and navigation system combinations within an airspace. RNP types are already being used by airspace planners to determine airspace utilization potential and as an input in defining route widths and traffic separation requirements.

The concept and application of RNP

RNP as a concept applies to navigation performance within an airspace and therefore affects both the airspace and the aircraft. RNP is intended to characterize an airspace through a statement of the navigation performance accuracy to be achieved within the airspace. The RNP type is based on a navigation performance accuracy value which is expected to be achieved at least 95 per cent of the time by the population of aircraft operating within the airspace.

The development of the RNP concept takes into account that current aircraft navigation systems are capable of achieving a predictable level of navigation performance accuracy and that a more efficient use of available airspace can be realized on the basis of this navigation capability.

RNAV operations within the RNP concept It is anticipated that most aircraft operating in the future environment will carry some type of RNAV equipment. The carriage of RNAV equipment may even be required in some regions or States. ICAO guidance material therefore makes frequent reference to the use of RNAV equipment.

As explained earlier, RNAV equipment operates by automatically determining the aircraft position from one or more of a variety of inputs. Distances along and across track are computed to provide the estimated time to a selected waypoint, together with a continuous indication of steering guidance that may be used, for example, as in a horizontal situation indicator (HSI). In some States, accuracy requirements are now, or will soon be, such that RNAV equipment must be coupled or capable of being coupled to the autopilot.

RNAV operations within the RNP concept would permit flight in any airspace within prescribed accuracy tolerances without the need to fly directly over ground-based navigation facilities. The application of RNAV techniques in various parts of the world has already been shown to provide a number of advantages over more conventional forms of navigation and to provide a number of benefits, including:

- Establishment of more direct routes permitting a reduction in flight distances;

- Establishment of dual or parallel routes to accommodate a greater flow of enroute traffic;

* Establishment of bypass routes for aircraft overflying high density terminal areas;

* Establishment of alternatives or contingency routes on either a planned or an ad hoc basis;

* Establishment of optimum locations for holding patterns; and

* Reduction in the number of ground navigation facilities.

The potential also exists to utilize RNP based on RNAV for the establishment of optimum arrival/departure routes and approaches; all of these benefits are advantageous to both ATS providers and users.

The provision of services and facilities for RNP Since RNP is defined by a statement of navigation performance accuracy, there is an obligation on the part of the States providing ATS service and also of the aircraft operator to provide the necessary equipment to achieve the RNP accuracy.

The ATS provider State must ensure that services (i.e. communications, navigation and surveillance (CNS)) within a given airspace will be adequate to provide safe separation based on a defined set of separation standards. The aircraft operator (and their States of Registry) must in turn ensure that the aircraft intending to operate in a specified RNP airspace is equipped to achieve the required navigation performance. It should be noted that compliance with RNP requirements can be achieved in many different ways and neither the State nor the aircraft operator is restricted as to how RNP is to be achieved, as long as it can be demonstrated that the requirements will be met.

General provisions of RNP

The implementation of RNP allows enhancement of ATS system capacity and efficiency while at the same time retaining or improving established system safety. The types of RNP were developed to provide known levels of accuracy for navigation and to support planning for the development of airspace designs, ATC and operational procedures.

Containment probability RNP types for enroute operations are established according to navigation performance accuracy in the horizontal plane, that is, lateral and longitudinal position fixing. In order to facilitate the use of RNP in airspace planning, this accuracy is expressed as a single parameter (the containment value). The containment value is the distance from the intended position within which flights would be found for at least 95 per cent of the total flying time. It is not possible to quantify the maximum distance which traffic is likely to deviate beyond this defined airspace.

RNP types In order to simplify RNP types and to make the required accuracy readily apparent to airspace planners, aircraft manufacturers and operators, the RNP type is specified by the accuracy value associated with the RNP airspace. As an example, RNP 1 may have a navigation performance accuracy of 1 mile or possibly less (1.85km), i.e. within the designated airspace, the navigation performance of the aircraft population is 1 mile (1.85km) on a 95 per cent containment basis.

RNP 1 is envisaged as supporting the most efficient ATS route operations by providing the most accurate position information through the use of RNAV, allowing the greatest flexibility in routing, routing changes and real time response to system needs. This classification also provides the most effective support for operations, procedures and airspace management for transition to and from an aerodrome to the required ATS route.

RNP 4 will support ATS routes and airspace design based on limited distance between navaids. This RNP type is normally associated with continental airspace.

RNP 12.6 will support limited optimized routing in areas with a reduced level of navigation facilities.

RNP 20 describes the minimum capability considered acceptable to support ATS route operations. This minimum level of performance is expected to be met by any aircraft in any controlled airspace at any time. Airspace, operations or procedures based on capabilities less than those of RNP 20 would not be implemented except in special circumstances.

Time frame for RNP implementation The primary means of achieving RNP is by the use of RNAV equipment which is already in widespread use. Many States and regions are developing considerable experience in such aspects of RNAV operations as airworthiness and operational approvals, airspace planning, aircraft separation and route spacing requirements, user techniques, training, publicity and information exchange.

Airspace and RNP

RNP may be applied to ATS routes, to an area or a volume of airspace, or any other airspace of defined dimensions. Additionally, when approved by the State or the appropriate ATC authority, unpublished tracks (i.e. random tracks) may be flight planned within designated and published RNP areas.

Defining RNP airspace An RNP type would be selected in order to meet requirements such as forecast traffic demand in a given airspace. This required navigation performance would determine the necessary level of aircraft equipage and airspace infrastructure.

Applying RNP in an airspace Ideally, airspace should have a single RNP type. However, RNP types may be mixed within a given airspace.

RNP can apply from takeoff to landing with the different phases of flight requiring different RNP types. As an example, an RNP type for takeoff and landing may be very stringent whereas the RNP type for enroute may be less demanding.

Relation of RNP to separation minima RNP is a navigation requirement and is only one factor to be used in the determination of required separation minima. RNP alone cannot and should not imply or express any separation standard or minima. Before any State makes a decision to establish route spacing and aircraft separation minima, the State must also consider the airspace infrastructure which includes surveillance and communications. In addition, the State must take into account other parameters such as intervention capability, capacity, airspace structure and occupancy or passing frequency. A general methodology for determining separation minima has been developed by the RGCSP. In general, however, the more stringent RNP types will support reduced separation standards leading to increases in airspace capacity.

RNP coordinate system As navigation systems evolve from station-referenced to earth-referenced, an important consideration is the geodetic datum used for determination of actual position.

Geodetic datums are used to establish the precise geographic position and elevation of features on the surface of the earth. They are established at various levels of administration (international, national and local) and form the legal basis for all positioning and navigation.

At present, there are many geodetic reference systems in use throughout the world which result in different latitude/longitude definitions of the same point on the ground, according to which system is used. Differences of several hundreds of metres are apparent in some areas of the world and the implications for aircraft flying under RNP conditions are such that errors of this magnitude may not always be tolerated, especially in terminal areas. Moreover, specific problems may also arise in enroute operations, for example, when aircraft are transferred between area control centres of adjacent countries where different geodetic reference datums are in use. Similarly, aircraft FMS software could employ a different geodetic reference datum from that used in a given area to locate ground-based navigation aids (e.g. DME), or earth-referenced navigation aids such as the GNSS. Flight test trials have attributed significant errors to the use of different geodetic reference datums in simulated high precision RNP environments.

Based on the difficulties associated with the widespread use of different geodetic datums, and the need to have a single datum for an integrated, satellite based CNS/ATM system, ICAO has chosen World Geodetic System (WGS)-84 as the common world geodetic datum as there is a need to:

* Convert coordinates of airport key positions and ground-based navigation aids to a common geodetic reference datum;

* Ensure that all such locations are surveyed to a common standard which provides optimum accuracy, such as that obtained by GNSS surveying techniques; and

* Ensure that all FMS software is referenced to a common geodetic datum.

The ultimate responsibility for the accuracy of position data for aviation use rests with States; however, a collective effort will be required to implement WGS-84 on a global basis before earth-referenced systems can be adopted for all classes of air navigation. States are now required to update their geodetic reference datums to WGS-84 by 1997.

Required navigation performance operations

Provision of navigation services in RNP airspace Since required levels of navigation performance will vary from area to area depending on traffic density and complexity of the tracks flown, States will have an obligation to define an RNP type of their airspace(s) to ensure that aircraft are navigated to the degree of accuracy required for ATC. States will be responsible for ensuring that sufficient navaids are provided and available to achieve the chosen RNP type and for providing the relevant information to operators. Providers of ATS therefore must also consider the parameters of the navigation aids they provide.

The levels of sophistication of communications, navigation and surveillance vary widely throughout the world. In turn, ATC separation minima which are used to safely separate aircraft operating within a specified area are dependent on these capabilities within the airspace. In establishing an RNP airspace or route, it will be necessary to define the separation minima or minimum protected airspace that applies. The RGCSP is developing a methodology to interrelate communications, navigation and surveillance, traffic density and other parameters in order to develop separation minima in given airspaces.

Provision of ATC for RNP airspace From an ATC point of view, it is considered that existing ATC techniques and equipment can continue to be used for RNP fixed or contingency ATS routes.

It is possible that closely spaced parallel tracks will be introduced, or routes will be established close to airspace currently reserved for other purposes. In such cases, some form of alert in case of track deviation or conflict may be necessary.

In the case of applying random tracks in RNP areas, an increasing need for changes to the ATC system will arise as follows:

* In areas of low traffic density the amount of change may be small, but account will have to be taken of flight plan processing, conflict detection and resolution;

* In areas of higher traffic density, ATC computer systems will have to accept and process flight plan data concerning random navigation. Air traffic controllers must be able to easily amend and update the relevant flight plan information in the computer system. Prediction and display of potential conflicts at the planning stage may be required; and

* Radar control may also require conflict alert and resolution, including selectable presentation of track prediction. ATC will require a method of showing the latitude and longitude of key crossing points on the predicted track. This might simply be displayed in terms of position in relation to a grid or by automatic readout of the latitude and longitude or name code.

The introduction of RNP areas including random tracks may bring about changes to the operation of ATC which would make it essential that additional training be provided, taking into account matters such as:

* Potentially different RNP type routes in the same sector;
* Transition between different RNP type areas;
* Radiotelephony (RTF) procedures;
* Revised military/civil and civil/civil coordination procedures;
* Conflict prediction and resolution along unpublished tracks; and
* Revised contingency procedures.

As more sophisticated navigation applications become more widely used (e.g. parallel offset capability, RNAV standard instrument departures (SID), and standard instrument arrivals (STAR), holding and approaches), their integration into ATC procedures will require that controllers are trained to accept and exploit the use of these advanced capabilities.

Conclusions on RNP

RNP defines the capability for an aircraft to navigate in a particular airspace segment and allows the aircraft operator the choice of the specific equipment to achieve that capability. In the CNS/ATM environment, it is expected that the required capability could be provided by GNSS which will be able to provide a high integrity, highly accurate navigation service, suitable for sole means navigation for enroute, terminal and non-precision and precision landing operations.

Summary of the benefits of the future navigation systems

* The global navigation satellite system will provide a high integrity, high accuracy, worldwide navigation service for the enroute,

terminal, and non-precision approach phases of flight, and eventually for Category I, II and III precision approach and landing operations (ICAO, 1993a). GNSS will also make it possible to achieve capacity improvements at limited cost throughout the world;

* Aircraft will be able to navigate in all airspace in any part of the world using a single set of navigation avionics;

* Three and four dimensional navigational accuracy will be improved;

* Provider States of ATS will realize cost savings as existing ground-based navigation aids are no longer needed; and,

* GNSS systems can be used in conjunction with other systems, such as INS, to support RNP. Figure 4.4 displays the differences between the present and future navigation systems.

A glance at the future

> There is no question that GPS will be the navigation and landing system of tomorrow. You will see rapid implementation of GPS, faster than any other technology has ever been accepted (Del Balzo, 1993, p. 16).

Additionally, Del Balzo, the acting FAA Administrator at the time, called GNSS a "technology revolution" that could, when combined with other technologies such as data link, save as much as $5 billion per year and also help to eliminate ATS disparities around the world.

Almost everyone agrees that satellite based navigation, especially when integrated with CNS/ATM systems, has enormous technical and savings potential. There are several trials and demonstration projects taking place

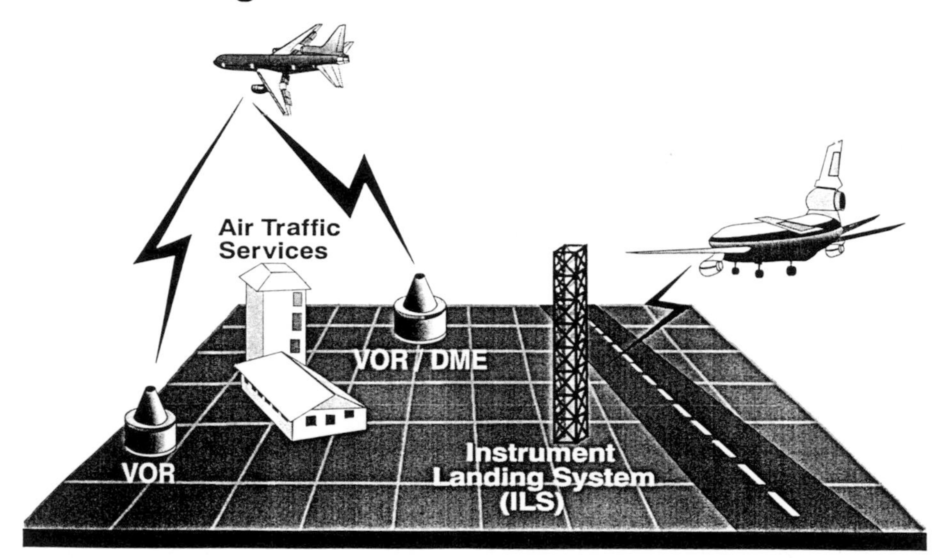

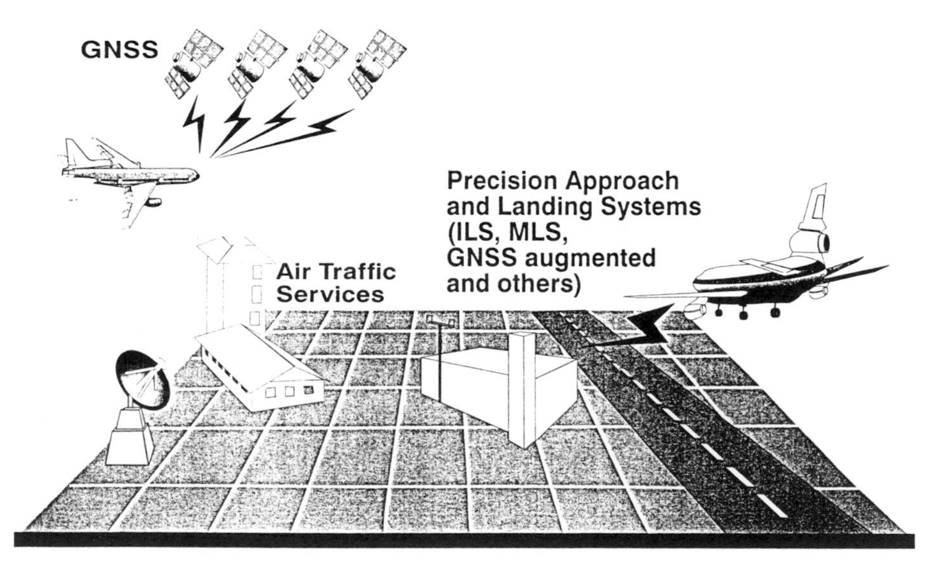

Figure 4.4 Differences between the present and future navigation systems

around the world, carried out by manufacturers, States and universities, and mostly sponsored by States.

In the civil aviation sector, industry and government driven projects are generating new capabilities that enhance the accuracy of GNSS both on a local and broad scale, enabling it to be used as a sole means of navigation.

Some of the latest developments concerning GNSS are outlined in the following chapters.

Use of GPS in the national airspace system of the United States

As one of the only two owner and provider States of GNSS, the United States stands to gain immensely from early and substantial use of GNSS for civil aviation. This fact has motivated the FAA of the United States to undertake an aggressive programme of trials and demonstration projects leading to implementation of GPS.

A phased implementation process has therefore begun the institutionalization of GPS into the United States' National Airspace System and demonstrates the strong commitment of the United States toward early implementation of satellite technology.

Use of GPS On June 9, 1993 the FAA announced approval for use of GPS in the oceanic, domestic enroute, terminal and non-precision approach phases of flight. This marked the first operational authorization to use GPS in instrument flight rules (IFR) conditions using avionics certified according to current United States technical standards. Also included in the approval were provisions for properly certified avionics to be used to fly all existing non-precision approaches (except localizer based) at more than 2,500 airports. Nearly 5,000 GPS approaches therefore, became immediately available for GPS based use (FAA, 1994a).

In 1994, the FAA approved the use of GPS as a primary means of navigation for over-the-ocean and remote aircraft operations. That move came three months ahead of the previously planned date. Primary GPS use in domestic airspace was scheduled for 1997 (Norris, 1994b).

GPS specific approach procedure design criteria are also in place to develop what is estimated to be another 10,000 procedures in the future for qualified airports currently not serviced by ground navigation aids.

The approval process for integrating GPS into use as an approach aid for non-precision approaches has been divided into three phases, all requiring properly certified avionics. The first phase allows use of GPS with several constraints, the most critical being the requirement for active monitoring of the appropriate ground navigation aid supporting the approach, to provide integrity.

The second phase began with the declaration of Initial Operational Capability (IOC) of GPS. On February 17, 1994 the Administrator of the FAA announced that GPS was available for civil use, thereby making it an integrated part of the United States National Airspace System (NAS).

Now that a fully operational constellation is in place, active monitoring of the ground navigation aid supporting a particular airport, as was indicated in the first phase, is no longer required, as integrity is now provided by Receiver Autonomous Integrity Monitoring (RAIM).

Phase three involves the actual change of procedure names to include GPS in the title of the approach. The most significant change concerning phase three however, is that the ground navigation aid supporting the approach at the destination airport is not required to be operating.

Research and development The FAA research and development programme for GPS targets several techniques aimed at satisfying primary means of navigation requirements for all phases of flight down to category III precision approaches. The FAA maintains an aggressive GPS research and development programme which is closely tied to the parallel development of operational requirements. The FAA believes that international involvement has been a key to the success of its programme as indicated by significant work accomplished with Canada, Fiji and Eurocontrol (FAA, 1994b).

Wide Area Augmentation System (WAAS) One area where the FAA continues to move forward at a rapid rate is that of the WAAS. The WAAS will consist of geostationary satellites and an independent network of monitoring stations to augment GPS for civil aviation as a primary means navigation aid from enroute through to precision approaches. The WAAS will enhance the broadcast of a GPS-like ranging signal and provide integrity broadcasts to permit aviation users to determine when GPS should not be used for each of the various phases of flight. WAAS will offer a vertical accuracy of four to seven meters, close to that required for Category I approaches (Warwick, 1995).

WAAS works by receiving and processing data from GPS satellites at widely dispersed sites referred to as wide area reference stations (WRS).

The data is forwarded to processing sites known as wide area master stations (WMS), which process the data to determine the integrity for each monitored satellite and then generate a navigation message to be transmitted to geostationary satellites (GEO). This is then sent to suitably located ground earth stations (GES) and uplinked along with the GEO navigation message to the GEO satellites. These GEO satellites then downlink this data on the GPS link frequency with a modulation similar to that used by GPS allowing users to have this correction information readily available.

In addition to providing GPS integrity, the WAAS will verify its own integrity and take any necessary action to ensure that the system meets WAAS defined performance requirements.

The FAA validated the concept of the WAAS by a cross country demonstration flight in December 1993. This was considered as a milestone in the use of GPS, demonstrating the use of a network of eight ground stations strategically placed across the U.S. As an initial demonstration, this test showed that coverage could be provided through the use of ground monitor stations and geostationary satellites.

The FAA plans to accelerate the implementation of GPS for enroute, through to precision approach using the WAAS (FAA, 1994b). The implementation will be divided into two phases; Phase One: initial operating system; and Phase Two: complete system (primary means capability). The WAAS is expected to be approved for supplemental use in 1998 and primary means use in about 2002. The end-state WAAS will consist of eight geostationary satellites which, together with 24 GPS satellites, will provide a total of 32. Ground installations for the WAAS have been surveyed and Canada will contribute five and Mexico three to the total WAAS system. Additionally, an alternate Master Control facility is being planned to be installed (Peterson, 1995).

Infrastructure of the WAAS will include 24 wide area reference stations collocated with existing FAA air route traffic control centres and other facilities.

Local Area Augmentation Systems (LAAS) Local area differential research is focusing on early Category I precision approach capability for private users. Current activities in this area are being conducted with regional air carriers, which will be among the first to benefit from this technology. Work performed by the FAA demonstrated with high statistical confidence that code differential local area systems will perform well within requirements of Category I and near Category II systems.

Furthermore, the FAA is in the process of demonstrating the feasibility (accuracy and integrity) of DGPS based category II and/or III approaches and landings. These plans are moving along in line with FAA requirements and recent achievements by industry, government laboratories and academic institutions.

Flight tests using DGPS have demonstrated better than Category III lateral and longitudinal accuracy and better than Category I vertical accuracy. The research is therefore now structured to draw upon the best efforts of industry, government and academia to demonstrate Category III vertical accuracy and sufficient system integrity. To this end, industry proposals continue to be solicited, evaluated and tested. A cooperative agreement has been signed with the National Aeronautics and Space Administration (NASA) of the United States to do similar work. Several universities and manufacturers are being sponsored, in addition to many independent efforts, aimed at development and demonstration of innovative techniques for DGPS precision approaches. Based on the results achieved so far, the FAA has concluded that differential, or now more commonly known as augmented, GPS can meet Category III requirements consistently and is also technically feasible. The FAA acknowledges, however, that for an operational system, additional investigation is still needed in the areas of integrity, availability, continuity and certification (ICAO, 1995f; CNS Outlook, 1995) and that commercially available GPS based Category III landing systems will not be on the market before the year 2000 (Peterson, 1995) or later.

United States' offer of GPS to be used by the international civil aviation community Based upon the recommendation of the FANS Committee, and to further the development of CNS/ATM systems, the United States has made available the GPS Standard Positioning Signal (SPS) for civil aviation use (ICAO, 1991a). This offer was made at the ICAO Tenth Air Navigation Conference in September 1991 in response to a formal request by ICAO. In response to several questions concerning the offer, the United States extended the scope of the 1991 offer at the 29th ICAO Assembly in September 1992 (FAA, 1994c). At that time, the United States further stated its intention to provide GPS-SPS for the foreseeable future and to also provide a minimum of six years advance notice of termination of GPS-SPS. In addition, this service has been offered free of direct user charges for the foreseeable future (Pozesky, 1993).

In April of 1994, the FAA Administrator reiterated the offer made at the 29th Assembly in formal correspondence with the President of the ICAO

Council. This correspondence represents a commitment on the part of the United States regarding the use of GPS-SPS by international civil aviation. (The offer by the United States, and the acceptance by the President of the ICAO Council are reproduced at Appendix H.)

Conclusion on the United States GPS programme

On 29 March 1996, President Clinton signed a Presidential Decision Directive (PDD) announcing a comprehensive national policy on the management and use of the United States GPS and related augmentation system. The PDD states that the United States will terminate the current practice of degrading civil GPS signals within the next decade.

> The United States is making significant progress in the exploitation of GPS technology for civil application. Working with industry and international agencies, the FAA has been successful in leveraging government research and development funds with outside resources to provide a robust and dynamic development programme capable of addressing a myriad of technical areas in parallel, delivering the most complete product to the flying public in minimum time (FAA, 1994b).

The Russian Federation

As pointed out in the previous chapter, the expansive airspace over Siberia and the Russian Far East has been closed to international civil aviation for many years. As a result of the break up of the Soviet Union and the political developments stemming therefrom, fifteen new States have now been created in the eastern part of Europe. Vast segments of previously restricted airspaces are now becoming available for international civil aviation. The only hinderance to a more healthy growth is a dilapidated or nonexistent air navigation infrastructure in this area.

It has always been known by the airlines that great savings in time and fuel could be achieved by making use of the airspace in the eastern part of Europe. Flights from the United States or Western Europe for example, to the rapidly growing Asian region, could fly more direct routings using shorter routes through this airspace. The Russian Federation and the other new States in the area are keenly aware of these possibilities and have been working closely with ICAO and neighbouring States toward opening new routes and developing the associated procedures and infrastructure.

For this reason, the European Air Navigation Planning Group (EANPG), the regional planning body made up of member States of the European Region, has begun the work of planning for CNS/ATM implementation in this area. Additionally, several arrangements outside of ICAO are pursuing the opening of this airspace. One such arrangement is the Russian American Coordinating Group on Air Traffic Control (RACGAT).

The Russian Federation is working actively for the early implementation and efficient use of CNS/ATM technologies. Not only does Russia stand to gain in direct payments from the airlines using their airspace, but in Russia alone over 400 newly established airlines had been registered by 1994, many of which continue to conduct operations (Russian Department of Air Transport, Russian Commission for Air Traffic Regulation and the U.S. Federal Aviation Administration, 1994). As the economies of the States in this region begin to grow, there will be a need for sufficient transportation to support the growth and an increasing need to travel for both business and pleasure.

Furthermore, as one of the two provider States of GNSS, Russia has gone to great lengths to bring their GLONASS system into operation and to assure the international community of its continued availability. V.Kuranov (1994), the Russian delegate to the GNSSP, informed the first meeting of that group that in 1993, Russia had completed State trials of the GLONASS system, fully confirming the assumed technical parameters and the stated characteristics of the system and, in accordance with a decision of the President of Russia, the system was adopted for operation with a space segment comprising twelve satellites at the time.

The Russian ministries of transport and defence have prepared a document entitled, *Regulations Concerning GLONASS System Use for Civil Users*. The document comprises legal aspects and guarantees necessary to the civil community. It also provides the users with necessary information on system status, outlining the steps being taken to ensure that civil users' interests were being taken into account in respect of system development and modernization.

Integration of GPS and GLONASS At the ICAO Tenth Air Navigation Conference (1991a), the results of a report prepared by the Massachusetts Institute of Technology (MIT) Lincoln Laboratory were presented, which indicated that GPS and GLONASS together would be capable of providing adequate redundancy to allow for system integrity monitoring (detection and identification of a failure in the system). Coverage provided by the two systems together appeared to be adequate for receiver autonomous integrity monitoring (RAIM) to be practical.

You will recall the discussion on integrity earlier in this chapter. Two types of integrity have been identified as *supplemental* and *sole means*. Supplemental integrity refers to the ability of the system to provide timely warning when the system should not be used for navigation. This means that the system must be able to detect a satellite failure before the error in the navigation solution exceeds the alarm limit identified for each phase of flight. Sole means integrity, being the more stringent of the two, refers to the ability of the system to remove erroneous satellites from the solution (isolate the satellite) before the error in the navigation solution exceeds the alarm limit.

To meet the growing demand for GNSS service, Northwest Airlines, Honeywell, the Leningrad Scientific Research Radio Technical Institute (LSRRI) and the All Union Scientific Research Institute of Radio Equipment (AUSRIRE) entered into an agreement (Hartman, 1992) to investigate the capabilities and limitations of integrating the signals from the GPS and GLONASS satellites. The study consisted of both lab and flight tests and found that with full deployment of both systems, the two constellations would be very close in size, accuracy and availability; the benefits of an integrated receiver would be significant, and as the signal structure of the GLONASS satellite signal is similar to that of GPS, the use of both GPS and GLONASS satellite signals in a single integrated receiver design would be possible. It was found that the RAIM solution would be available 100 percent of the time, even after the failure of three GPS and three GLONASS satellites. Furthermore, since the GLONASS satellites would not have the intentional accuracy degradation of selective availability, as is the case with GPS, due to United States military requirements, the GLONASS satellites would have the potential of providing twenty five meter navigational accuracies.

With the expected widespread availability of GLONASS receivers, users will have the two independent sources of navigation signals. The study (Hartman, 1992) anticipated that these findings would increase the international acceptance of satellite navigation as a sole means worldwide navigation service.

Not awaiting the full deployment of GLONASS, nor availability of commercially developed combined GPS-GLONASS receivers, the Russian Federation successfully performed experimental Category I approaches by operating GPS-GLONASS combined systems in differential mode. Experiments are also being conducted for Category II and III approaches.

Taking into account previous ICAO recommendations on the combined use of GPS and GLONASS, the Russian Federation has, on several occasions, expressed its willingness to conclude necessary agreements in order to meet the requirements of the international civil aviation community.

In March of 1995, the Russian Federation issued several decrees concerning the use of GLONASS for civil use. The first decree establishes the working procedures leading to the use of GLONASS for civil purposes. Following up on this, the Russian Ministry of Transport had been instructed to develop, by the second quarter of 1995, a coordinating council for the use of GLONASS by Russian and international civil users, leading to a GLONASS use programme encompassing development and production of civil navigation equipment for the period 1995-2000. The decrees also cover the areas of establishment of an information service concerning the GPS and GLONASS systems and the submission to ICAO of all relevant materials necessary for the preparation of an agreement on the use of GLONASS as an element of the international GNSS for civil use.

European Commission

At a meeting of the U.S. Coast Guard Civil GPS Service Interface Committee (CGSIC) in August 1995, a representative of the European Commission announced that the Commission's 15 member States had issued a full endorsement of GNSS for multimodal navigation use in Europe. The Commission sees Europe's contribution to GNSS as taking the form of completing and operating the European Geostationary Overlay Service (EGNOS), a GNSS augmentation system similar to the FAA WAAS. EGNOS is expected to use both GPS and GLONASS, as well as certain geostationary satellites. It is expected that as part of the project, EGNOS and WAAS will eventually become linked (CNS Outlook, 1995).

Australia

Australia has developed a transition programme to GNSS under the guidelines of the Asia/Pacific Regional Implementation Plan for new CNS/ATM systems, developed by the ICAO Asia/Pacific Air Navigation Planning and Implementation Regional Group (APANPIRG) (see Chapter 2 for a description of ICAO regional planning groups).

Australia has stated their recognition that operational advantages can be gained with early transition to satellite based navigation and is actively pursuing its early operational acceptance and use (Howell, 1994).

The Australian plan foresees the use of satellite based navigation as occurring in three distinct phases. Phase One would encompass limited approvals, based upon use of the existing GPS and GLONASS systems; Phase Two would see the deployment of a wide area augmentation system; and Phase Three would realize the introduction of local area augmented systems.

Canada

Canada has stated that it is satisfied with the assurances of the United States Government that GPS will be made available for everyone, including civil aviation users, free of charge for the foreseeable future, and that they view GPS no differently in these respects than any other government navigation aid. This is a significant statement in light of the thorny institutional issues that remain to be solved. Canada has further stated that it is committed to the development and implementation of satellite navigation (ICAO, 1994h) and is also participating with the United States in the WAAS programme through the contribution of several ground stations.

Supplemental use of GNSS for enroute navigation in Canadian airspace, both domestically and over oceanic airspace as well as for non-precision approaches was approved in July 1993 with certain requirements and limitations.

Additionally, Canada continues to carry out an aggressive work programme aimed at, among other things, the progressive reduction in dependency on ground based navigation aids.

Fiji

In any discussion of CNS/ATM, the State of Fiji deserves a special examination. Fiji occupies a unique central geographical position in the South West Pacific Ocean. It is a country of 300 small islands with a population of 736,000 and is responsible for a flight information region (FIR) of 7.5 million square kilometres, straddling the major air routes between Australia and the United States (Yee, 1994). Mr. Yee, in his presentation to the GNSS Panel noted that Fiji only became involved in CNS/ATM activities in 1990, but from the outset, Fiji had no doubts as to the "enormous benefits promised by the FANS concept and today, the promises of huge cost savings, fuel efficient tracks, time savings and improved safety are no longer promises but a hard reality.... For this reason, Fiji is playing an active part in the implementation of CNS/ATM in the Pacific Region" (1994, p.2) .

Fiji's most exciting project, according to Yee, involves the use of GPS for navigation in the domestic environment instead of continuing to rely on ground based navigation aids. The scattered nature of Fiji's many islands, the six months of rainy season which makes visual flight rules (VFR) navigation very difficult, and the lack of ground aids, have made GPS the ideal solution to their problems.

After eighteen months of planning and seven months of successful trials and demonstrations, Fiji commenced using GPS as the primary navigation aid for enroute and terminal operations on 15 April 1994, becoming the first country in the world to do so. Fiji expects to eventually use GPS for approaches after the receivers are upgraded to handle this aspect of navigation.

Fiji estimates that it would have cost them 100,000 U.S. Dollars to install and maintain one non-directional radio beacon (NDB). This is the amount paid for all of Fiji's GPS receivers. They further estimate that 2 million U.S. Dollars have been saved by not installing NDBs and improved airline efficiency saves 12,000 U.S. Dollars per annum.

Based on their initial success, charts are being drawn to allow aircraft to make instrument approaches to all airports under visual conditions using GPS. Additionally, Fiji is installing an augmented GPS system at their main international airport in parallel with its instrument landing system (ILS), which is due for replacement in 1996 at a cost of 600,000 U.S. Dollars. It is expected that replacement of the ILS will be accomplished by the DGPS at approximately 120,000 U.S. Dollars.

For enroute oceanic, Fiji will take full advantage of the FANS 1 package described in Chapter 3, continuing to make use of GPS for navigation purposes.

In pursuing its goal of the most cost effective and efficient air navigation possible, Fiji is experimenting with using a flight planning system to track GPS fitted aircraft on its domestic network using VHF datalink and hopes to avoid having to install a radar system. Surveillance is covered in the next chapter.

There is no reason why the Fiji success story cannot also be the success story of every State and Region, regardless of their operational environment. And, by adopting a strategy of developing guidance material that States and Regions could use as guides for realizing early benefits from satellite navigation, we could also ensure that the paths they take remain consistent with the end state vision (Dorfler, 1994).

Germany

The German air navigation services agency, the DFS, has been conducting a satellite navigation test programme which could lead to satellite based non-precision approaches being allowed before the end of 1995 (Flight International, 1995). The main goal is to prove the safe use of satellite navigation in non-precision approaches and to examine possible disturbances in the reception of satellite signals.

The DFS has also demonstrated, that under certain conditions, augmented GPS can achieve the accuracies required for near category I precision approach minima.

Norway

Norway is developing a system that will provide accurate augmented GPS navigation throughout that country for air, land and sea users, by broadcasting differential error corrections over its AM and FM broadcast stations (Nordwall, 1993).

The satellite based reference system (SATREF) will give real time five meter navigation accuracy to mobile users and centimetre accuracy to stationery receivers. Each reference system compares its known location with its GPS derived position to determine errors in the satellite data, which it then sends to the control centre. Those error corrections will be valid for any other GPS receivers within several hundred kilometres. The control centre monitors the GPS data and controls the quality of data sent to the transmitters. A DGPS correction signal is broadcast on a sideband of an existing AM or FM signal and is undetectable by radio listeners. Some equipment is required by the users to receive and convert the information into the GPS receiver format.

Saudi Arabia

Being keenly aware of the difficulties involved with maintaining a navigational infrastructure in the inhospitable desert areas of Saudi Arabia, and realizing that CNS/ATM could be particularly beneficial in this respect, the Government of Saudi Arabia embarked on a trial programme in 1993 in order to gain experience and to allow the users of their airspace the opportunity to participate in planning requirements (Abudaowd, 1993). Saudi Arabian National Airlines (Saudia), the largest airline in the Middle East, has expressed a particular interest in this study.

The experimental programme concerning navigation, consists of flight inspection aircraft being equipped with GPS receivers, and a differential GPS station installation on the ground at the international airport at Jeddah. The facility is being used to conduct trial GNSS precision approaches. A major study is also underway which aims at the possibility of a complete upgrade of Saudi Arabia's air navigation infrastructure based on CNS/ATM.

Sweden

An innovative research, development, trial and demonstration programme related to CNS/ATM technology is underway in Sweden, now known as the GNSS-synchronized, self-organizing, time-division, multiple-access (STDMA) data link. This effort focuses on the use of satellite technology and datalinks for advanced communication, navigation and surveillance applications. The datalink system is designed to create a common technical platform which supports multiple CNS applications, such as uplink of GNSS augmentation signals, output of GNSS navigation data to avionics equipment and broadcast of automatic dependent surveillance data (Westermark, 1994).

The system under experimentation by Sweden offers several possibilities concerning navigation that have particular relevance to wide and local area augmentation and may also offer potential to provide a fall back mode in case of GNSS receiver failure or outage of the GNSS space segment. One of the more interesting prospects for the system however, has to do with its ability to broadcast surveillance data based on GNSS navigation data. For this reason this system is described in more detail in the next chapter dealing with surveillance.

Inmarsat

As mentioned in the previous chapter, the Inmarsat constellation of satellites has been recognized by ICAO as an important and available resource for early and continued implementation of CNS/ATM technology. In the case of navigation, Inmarsat has provisionally allocated navigation transponders on its next generation of satellites (Inmarsat-3). The allocation of these transponders was awarded to Inmarsat signatories (see previous chapter for an explanation of Inmarsat) who then make the transponders available to service providers in their respective States (Aeronautical Satellite News [ASA], 1995).

The five Inmarsat-3 satellites to be launched from late 1995 or early 1996, will each carry one navigation transponder which will be used to provide a civil component to GPS and GLONASS signals. These satellites may also form the first elements in an evolving international civil GNSS (ICAO, 1995e).

Data provided by the Inmarsat payloads will allow satellite navigation to meet the stringent requirements of the civil aviation community related to reliability, availability and integrity of information provided by all satellites associated with GNSS (ASA, 1994a). The Inmarsat navigation package will be used to provide four different services (Ryan, 1994) as follows:

Integrity monitoring The information from GPS and GLONASS satellites contain information on the status of their transmissions as perceived by the control centres. With some failure conditions there can be a delay of as much as thirty minutes between the occurrence of the failure and the amending of the message being transmitted by the satellite to warn the user. Using the navigation package on Inmarsat-3, the status of all GPS and GLONASS satellites could be broadcast directly to users in essentially real-time. A network of ground stations monitoring GPS/GLONASS signals would provide the data.

Additional ranging source Under certain conditions, suitably modified GPS receivers can acquire and use retransmitted GPS signals. Very broadly, each Inmarsat satellite transponder will receive navigation related signals from a navigation station on the ground and retransmit these signals at power levels designed to produce approximately the same radio frequency levels as that of a GPS satellite. The signal would therefore be treated as a GPS signal by any GPS receiver. The geographical coverage and hence the availability of GPS would therefore be dramatically improved since each Inmarsat-3 transponder, being on geostationary satellites, at any given time, can see the same area as three orbiting GPS/GLONASS satellites.

Accuracy enhancements The Inmarsat navigation package may also be used to transmit wide area differential correction messages to aircraft, as will be the case with the WAAS being planned by the United States. It is estimated that an accuracy of eight meters (99.9 per cent) can be achieved using wide area differential corrections at any point within a global beam.

Accurate time The Inmarsat navigation package could also be used to provide extremely accurate time references. These could make communication systems much more efficient and could also be used as a common reference for the ATC systems of the world.

Access to Inmarsat navigation transponders would be available through navigation earth stations operated by Inmarsat signatories. Signatories may also provide the ranging signal. The integrity signals and, if provided, the wide area differential signals, will be the responsibility of the service provider who will deliver the signal for transmission over the transponder to the navigation earth station.

Additionally

* Stanford University scientists have developed another technique that could prove sufficiently accurate for Category III instrument approaches. Their Kinematic GPS landing system extends the accuracy of D-GPS by tracking the carrier phase of the GPS signal. Data calculated measure the aircraft's position with vertical accuracies of one foot (Klass, 1993);

* The 29th Session of the Assembly of ICAO urged the development of provisions and guidance material relating to all aspects of CNS/ATM through the convening of adequate meetings, conferences, panels and workshops with the participation of Contracting States. In line with this, yearly symposia have now been established by the International Air Transport Association (IATA) in cooperation with ICAO, known as Global NavCom. Global NavCom has established itself as an international forum for the discussion of FANS CNS/ATM implementation issues (Rochat,

1995). Over 500 world aviation leaders attended the first Global NavCom in 1993 to launch what they believed to be "the start of a new era in air traffic management" (Olsen, 1993);

* The first meeting of the ICAO GNSSP was held in Montreal from 17 to 25 October 1994. Twenty three panel members, fifty advisers as well as six observers from twenty States and eleven international organizations participated in the meeting (see Chapter 2 for a general description of panels). That meeting considered performance criteria for application of GNSS, including augmentation subsystems and integrity monitoring and identification of institutional implications (ICAO, 1994e). The meeting established two working groups. One group is developing the materials that States could use as guidelines for providing a means to realize early operational benefits from existing systems. The second group is developing the SARPs, guidance material and system validation requirements that need to be developed to meet the established RNP requirements (Airport Forum, 1995);

* A document that offers guidelines for realizing early benefits from existing satellite based navigation systems was circulated to States by ICAO. It was agreed that this document would become an ICAO Circular for public sale;

* The second meeting of the GNSSP was held in November of 1995 and work continued on the development of SARPs and guidance material;

* A meeting of an AWOP Working Group took place in late 1995 where work concentrated on the application of GNSS for approach, landing, departure and surface movement operations and development of RNP for these operations.

Recapping the major points

In modern times, there have been several methods developed for navigation, making use of electronic means and instruments with great success around coastal waters and over land. An exact system for navigating reliably over all parts of the globe did not come into man's reach until very recently. That is because new methods of determining one's position, regardless of location, had to await the introduction and advanced use of satellite and computer technology. ICAO has given the generic term of Global Navigation Satellite System to the concept of navigation by satellite encompassing satellite constellations, aircraft receivers, and system integrity monitoring.

This chapter focused on the envisioned changes that will form the basis of future navigation systems and methods and discussed some of the benefits that would come to ATM as a result of GNSS.

The concept of RNP which will define the performance required in a particular airspace or phase of flight, was examined in some detail as it will have a major impact on the planning and implementation requirements connected with future satellite navigation systems.

Finally

* GNSS will provide worldwide navigational coverage and will be used for aircraft navigation, and also for non-precision approaches at first and later for precision approaches;

* To enable flexibility and to support the development of more flexible route systems and an RNAV environment, the concept of RNP has been developed. This concept allows a required navigational performance to be achieved by a variety of means;

* Area navigation (RNAV) capability will be progressively introduced in compliance with RNP criteria;

* Augmentation systems, which allow differential corrections to be made and transmitted to aircraft, will eventually provide highly reliable, highly accurate and high integrity global coverage and will eventually meet the navigation systems' requirements for sole means of navigation for civil aviation;

* The present navigation system serving enroute navigation and non-precision approaches will be able to meet the RNP conditions and coexist with satellite navigation systems, however, it is foreseen that satellite systems will eventually become the sole means of radio navigation allowing the progressive withdrawal of ground-based systems, including instrument landing system (ILS) (ICAO, 1991a).

Questions and exercises to expand your knowledge

1) With current ATC systems, aircraft usually follow predetermined airways or routes for a few different reasons as follows:

   * Present navigation methods often require that aircraft fly over ground based navaids;

   * Air traffic controllers are not capable of separating aircraft from each other if they all operate on random routes.

   How will the use of GNSS and the required navigation performance (RNP) concept address the problems mentioned above?

2) There is a wealth of information available on how GPS or GNSS in generally works. Explain how GNSS receiving equipment determines its position.

3) What is the difference between "sole means" and "supplemental means" of navigation?

4) In the context of GNSS, what does "integrity" refer to?

5) How can problems associated with integrity be addressed?

6) Explain how a wide area augmentation system (WAAS) works.

7) Draw a diagram of an aircraft on final approach to an airport, using GPS signals and local area differential correction messages from a D-GPS station to conduct the approach.

8) Explain the advantages and benefits to providers and users of ATC services that would be had with the introduction of the required navigation performance (RNP) concept.

9) Draw a diagram of an aircraft enroute, using satellites for navigation, also making use of a wide area augmentation system (WAAS). Among other things, you should depict the following in your diagram:

   * Navigation satellites;
   * Inmarsat satellites used for WAAS;
   * Ground reference station for WAAS;
   * ATC unit;
   * Aircraft.

10) This chapter makes reference to the World Geodetic System (WGS) 84. Presently, several different geodetic systems are used by different States and sometimes more than one are used internally within a State. Explain a geodetic reference system. Why is it critical that the world adopt one common reference system for the future CNS/ATM system?

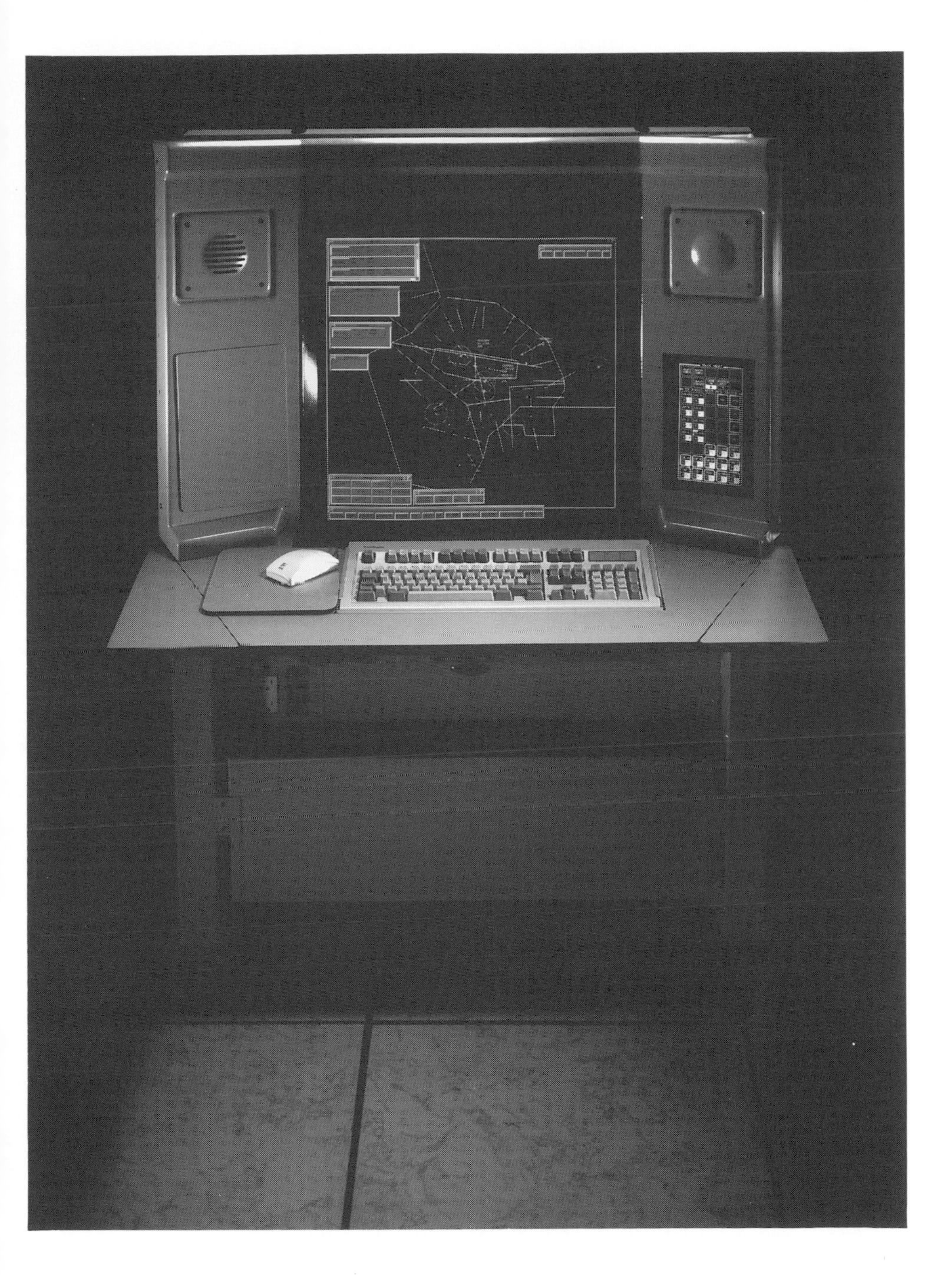

A common air traffic controller workstation for an area control centre (ACC), incorporating advanced software and human factors engineering principles. (Picture provided courtesy of Hughes.)

5. Surveillance

Introduction

> Timely provision of an accurate aircraft position and reliable communications are the keys to the operation of an effective ATC system (Castro-Rodriguez, 1990).

Previous chapters have shown how aircraft operating in future CNS/ATM systems will communicate with people and systems on the ground and how they will navigate. For effective air traffic management (ATM) to be possible however, people or systems on the ground must know the position of aircraft on a continuous basis and be able to estimate their future position.

Generally, the idea of keeping track of an aircraft's position is referred to as surveillance. Of course the most basic way of knowing an aircraft's position and of determining its future position is to communicate directly with the aircraft. In this way, a pilot would verbally inform the air traffic controller or air traffic services (ATS) provider of the aircraft's position and that information could, at the very least, be written on a strip of paper so that the aircraft could be tracked and/or separated from other aircraft. A more advanced form of surveillance however, and what is more commonly used in busy continental airspace, is based on the use of radar. The various types of radars will be briefly discussed in this chapter.

With the introduction of air ground data links as described in Chapter 3, together with sufficiently accurate and reliable aircraft navigation systems, as described in Chapter 4, a new method of surveillance known as automatic dependent surveillance (ADS) has evolved and will form an integral part of CNS/ATM systems.

ADS is a surveillance technique for use by ATS in which aircraft automatically provide, via data link, data derived from on-board navigation and position fixing systems. The data includes as a minimum, aircraft identification and four dimensional position (ICAO, 1995c). This information is then used to monitor progress of aircraft and to determine aircraft proximity in relation to other aircraft. In an automated ATM system, this information could be used to dynamically determine future potential conflictions with other aircraft and resolutions.

By now, you should begin to recognize the relationship between the various components of communications, navigation and surveillance, and how they are being combined, or soon will be, in new and innovative ways using advanced technologies, previously unavailable, to form an advanced and more efficient ATM system. You will certainly appreciate this as you learn more about the surveillance function.

As mentioned in Chapters 3 and 4, the use and integration of CNS/ATM systems will be different than it is today partly because of a much greater use of satellite technology. This is certainly true in the field of surveillance where ADS will be largely based on satellites.

This chapter provides background on the work of the FANS Committee and ICAO and its panels as that work is related to the surveillance component of CNS/ATM systems. This includes a description of the envisioned changes that will form the basis of future surveillance systems and methods, and a review of the benefits to aviation that could be expected with implementation of the new systems.

This review includes an analysis of the concepts of ADS and Mode S radar as these technologies will form an integral part of the CNS/ATM surveillance element.

This chapter also discusses some of the ways and places where the surveillance element of CNS/ATM technology is being experimented with and implemented.

Definitions

In order to fully understand this chapter, it is necessary to first become familiar with a few of the common international terms which are frequently used when referring to surveillance and which you will see often as you become involved with international civil aviation.

> *Automatic dependent surveillance (ADS)* is a function for use by ATS in which aircraft automatically transmit, via data link, data derived from on-board navigation systems. As a minimum, the data include aircraft identification and three-dimensional position. Additional data may be provided as appropriate.
>
> *ADS-based ATC system* In addition to the definition for ADS, the ADS-based ATC system also includes the capability to exchange messages between pilot and controller via data link and by voice for emergency and non-routine communications.
>
> *Mode S* An enhanced mode of secondary surveillance radar (SSR) that permits the selective interrogation of Mode S transponders, the two way exchange of digital data between Mode S ground stations and transponders, and also the interrogation of Mode A/C transponders.
>
> *Mode S data link* A means of performing an interchange of digital data through the use of Mode S ground stations and transponders in accordance with defined protocols.
>
> *Primary radar* A radar system which uses reflected radio signals.
>
> *Radar* A radio detection device which provides information on range, azimuth and/or elevation.
>
> *Radar blip* A generic term for the visual indication, in non-symbolic form, on a radar display of the position of an aircraft obtained by primary or secondary radar.
>
> *Radar display* An electronic display of radar derived information depicting the position and movement of aircraft.
>
> *Secondary Surveillance radar* A surveillance radar system which uses transmitters/receivers (interrogators) and transponders.

Transponder A receiver/transmitter which will generate a reply signal upon proper interrogation; the interrogation and reply being on different frequencies.

Shortcomings of the present system

The efficiency of ATC is dependent upon the availability of communications and surveillance capabilities. Because of the limitations of line-of-sight systems, flights operating outside of surveillance radar must be controlled on the basis of their flight plans, updated by position reports transmitted on either VHF radio, when within range, or HF radio, which has operational limitations (Castro-Rodriguez, 1990). The type of ATC employed in such situations is known as procedural control. Many flight information regions (FIRs) throughout the world rely on procedural systems for controlling air traffic, often with little or no automation support. This is particularly true in most oceanic regions where radar and VHF communications cannot be provided.

In addition to oceanic areas, radar and VHF radio are often not employed in low traffic density areas in continental airspaces because of low cost effectiveness. Moreover, there are no surveillance systems comparable to radar that could be employed in these regions (ICAO, 1990). The application of procedural techniques in these areas ensures an adequate level of safety, however, at the expense of optimal flight profiles and system capacity.

Continuing increases in international passenger travel and the need to reduce commercial air transport operating costs, requires more efficient ATC techniques. ADS as envisaged by the FANS Committee, has been identified as one of the ways of overcoming the shortcomings inherent in the present system.

The previous chapter described the shortcomings of the present CNS system as identified by the FANS Committee, which recognized that for an ideal worldwide air navigation system, the ultimate objective is to provide a cost effective and efficient system that would permit the flexible employment of all types of operations in as near four dimensional freedom as their capability would permit (ICAO, 1988b).

Surveillance

For effective ATC to be possible, people or systems on the ground must know the position of aircraft on a continuous basis and be able to estimate their future position. The idea of keeping track of an aircraft's position is referred to as surveillance. In the vicinity of an aerodrome, an air traffic controller could look out the window of the tower cab and see the aircraft. Beyond this, the most basic and oldest method of knowing an aircraft's position and of determining its future position is to communicate directly with the aircraft. In this way, a pilot would verbally inform the air traffic controller or ATS provider of the aircraft's position and that information could be written on a strip of paper so that the aircraft could be tracked and/or separated from other aircraft. This is still quite common over the oceans where other forms of surveillance are not yet available, and in less developed parts of the world where air traffic is not very dense. ATC based on position reports is known as procedural control.

A more advanced form of surveillance however, and what is more commonly used in busy continental airspace and in the vicinity of busy aerodromes, is radar. As the ATC system has developed throughout the world, radar has become the most important tool used by air traffic controllers for surveillance of aircraft and weather. Radar allows the position of an aircraft to be presented on a radar display, where an air traffic controller provides radar control. Radar control is preferable to non-radar or procedural control, however, radar equipment must be purchased and maintained, which is not always feasible or possible, especially where traffic is not dense enough, or where it is physically impracticable, such as over the ocean or in the desert.

Primary surveillance radar (PSR)

ATC radar in its simplest form, known as primary radar, provides the controller with a visual indication, on a cathode ray tube, of all radar echoes reflected from aircraft within line of sight of the ground based radar facility. The display presented to the controller provides information on the range and azimuth of reflected objects, including aircraft. Because primary radar equipment in no way relies on any action on the part of the pilot or aircraft, it is known as independent surveillance.

Secondary surveillance radar (SSR)

SSR is composed of a ground interrogator and airborne transponder equipment. The ground interrogator equipment is normally collocated with a primary radar so that targets provided by the primary radar and those provided by SSR could be presented simultaneously on the controller's radar display and, in automated systems, appear as one single target.

SSR differs from PSR in that it operates in a request-reply mode whereby the ground equipment sends out a signal which, in turn, triggers a reply signal transmitted from the aircraft transponder, rather than relying on reflected signal returns from the aircraft. Because of a required action on the part of the aircraft, namely, the need for a transponder to reply, SSR is considered as being a dependent type of surveillance. The FANS Committee considered SSR as being a type of cooperative dependent surveillance (CIS), a phrase developed by the FANS Committee, however, since its inception, very little work has been done in the way of expanding on the concept of CIS for several technical and economic reasons. For the purposes of this text, therefore, SSR will be considered in the context of its traditional categorization, which is that of dependent surveillance. Overall, SSR offers several advantages over PSR.

The term *mode* is used to describe the type of ground transmission or interrogation which is used. Aircraft transponders reply to interrogations with specific *codes*. Among other capabilities, SSR radar, along with appropriate processing equipment, allows altitude information to be displayed in numeric form adjacent to the aircraft blip on the radar display (ICAO, 1984). Additionally, significant operational improvements can be achieved through automatic processing and display of SSR derived data. Using such equipment, computer generated symbols can be provided and associated with the radar target which can depict aircraft identification, type, altitude and ground speed. This processed information can significantly reduce problems associated with identification, tracking, altitude verification, monitoring of aircraft speed and frequency congestion, resulting in reductions of controller workload associated with strip marking and coordination. Automation associated with ATM is more fully addressed in Chapters 6 and 8.

Surveillance as envisaged in the integrated global CNS/ATM environment

Mode S radar, which is also described in this chapter, will have a significant role to play in CNS/ATM systems, especially in high density continental airspace and in the vicinity of major airports.

The introduction of air ground data links, together with sufficiently accurate and reliable aircraft navigation systems, presents the opportunity to provide surveillance services in areas which lack such services in the present infrastructure, particularly in oceanic airspace and other areas where current systems prove difficult or impossible to implement.

The major element of the future surveillance component of CNS/ATM systems will be ADS, which allows aircraft to automatically transmit their position and other data such as manoeuvre intent, speed and weather, via satellite or other communication link, to an ATM unit. This ability will make previously unusable airspace, usable, while also allowing a more efficient use of other airspace where air traffic separation must be increased because of a lack of adequate surveillance capability. ADS is also envisaged as a backup to SSR and for providing surveillance for surface movement at airports.

In addition to areas which are at present devoid of traffic position information, other than pilot provided position reports, ADS will find beneficial application in high density areas, where it may serve as an adjunct and/or backup for SSR and thereby reduce the need for primary radar as a backup system.

As with current surveillance systems, the full application of ADS requires complementary two way pilot-controller data and/or voice communication, at least for emergency and non-routine information (ICAO, 1988b).

The FANS Committee developed and expanded upon the roles of the various existing and unfolding surveillance elements in a CNS/ATM environment (ICAO, 1988b) according to the groupings: *cooperative independent surveillance (CIS); independent surveillance*; *ground movement surveillance*; *detection and location of aircraft under emergency conditions*; *and, airborne collision avoidance systems (ACAS)*. The foreseen use and evolution of these elements is described below, except for CIS, as this concept has not been significantly expanded upon in follow up of the work of the FANS Committee and is not seriously being considered at this point in time because of technical and economic reasons.

Dependent surveillance

Secondary surveillance radar (SSR) SSR is in wide use in many parts of the world where terrestrial surveillance systems are appropriate. By enlarging SSR with Mode S, the selective address and data link capabilities will further enhance the beneficial role of SSR for surveillance purposes.

Independent surveillance

Primary radar Although the use of primary radar is already declining, its need will continue to exist in those airspaces where there is a mix of SSR equipped aircraft with non SSR equipped aircraft and a need to provide compatible services to both categories. However, as such circumstances are decreasing, the need for primary radar will decrease. The decrease will be further advanced by the introduction of ADS, at least in its role as an adjunct or backup for SSR. It is foreseen that SSR, through the Mode S system and its data link, and ADS, through satellite communication, will be of such high integrity that it will diminish the justification of primary radar for ATC for international civil aviation. It is recognized, of course, that primary radar for other purposes, including weather detection, will continue to be needed.

Ground movement surveillance

At very high activity airports, there will be a need for advanced systems for airport surface communication, navigation and surveillance. The term commonly used to describe systems that assist ATC operations on the airport surface is surface movement guidance and control systems (SMGCS). The newer systems based on CNS/ATM technologies will provide more capability than surface movement radar alone. It is increasingly being accepted that technologies such as Mode S data link, satellite navigation and ADS techniques could be used for ground movement in a way in which no additional avionics would be required. Since there is an increasing need for such systems, the FANS Committee encouraged their development.

Among other things, advanced SMGCS will provide the possibility of automatically guiding aircraft and other vehicles around the movement areas of airports, assisting in spacing between aircraft and alerting controllers and pilots when aircraft are potentially in conflict with each other on the movement area.

Detection and location of aircraft under emergency conditions

Aircraft under emergency conditions can be detected and an approximate position established using their last message and their last position as recorded from the surveillance system. In this respect, aeronautical mobile satellite service (AMSS) and ADS will extend that possibility to worldwide coverage. The transmission of signals by emergency location beacons, which begin operating after an accident, allows a more complete detection of the emergency, an actual location of the grounded aircraft to be determined and the rescue party to home in on the aircraft. A system known as COSPAS-SARSAT which uses satellite payloads on low altitude polar orbits, has demonstrated very effectively, the feasibility of locating beacons in all areas of the world covered by satellite ground stations.

Airborne collision avoidance systems (ACAS)

ACAS equipment on-board aircraft operates as an airborne SSR radar by interrogating the transponders of other aircraft on the same interrogation/reply frequencies and modes, in its vicinity. This gives the pilot of an aircraft equipped with ACAS, an indication on a display, of the heading and altitude of the other aircraft equipped with SSR transponders. ACAS therefore acts as an alerting and avoidance system. Software advances in the latest ACAS versions provide pilots with avoidance manoeuvres, either vertically or horizontally, to follow in the event that another aircraft is on a course that the ACAS interprets as threatening.

The FANS Committee agreed that ACAS can provide information about proximate traffic, particularly where air to ground communications cannot be monitored by aircraft crews, and can, as a last resort, prevent collisions in the event of large errors or undetected blunders. The FANS Committee felt that ACAS could not be considered as a substitute for ATS service, however, it could have an effect on separation standards, with regard to that portion of the separation required to protect against potential risk associated with large errors or blunders.

Automatic dependent surveillance (ADS)

ADS has been defined by the FANS Committee as a function for use by ATS whereby aircraft automatically transmit, via data link, data derived from on-board navigation systems, including aircraft identification and three-dimensional position. The introduction and operation of ADS will provide ATS units with the capability to extend automatic surveillance of aircraft beyond present radar coverage and to monitor air traffic operations over oceanic and land areas where procedural ATC is presently applied. It is foreseen that the implementation of ADS will have considerable impact on ATC procedures (ICAO, 1990).

A circular has been produced by the ICAO (1990) Secretariat with the assistance of an air navigation ADS Study Group (now the ADS Panel) comprising experts from ten Contracting States and five international organizations. It contains guidance material on ADS including information on the experience gained regarding development and planning of ADS, pre-implementation programmes and harmonization of ATM. Excerpts of the former ADS Study Group's work on ADS are reproduced below.

Principles of operation

The ADS concept is based on the use of digital data link (see Chapter 3 for a description of digital data link) communications and encompasses the transfer of the aircraft-derived position information to the controller in near real-time. This is accomplished automatically without the need for direct pilot or controller involvement. The information transferred comprises:

* Position information derived from the system being used to navigate the aircraft with a data rate which can be controlled by the ground ATC system;

* Information on the route stored in the system being used to navigate the aircraft, enabling the ATC system to verify the "next position" and the "following significant point" for conformance checking against the current flight plan (ATC clearance); and,

* Meteorological data for ATS and appropriate meteorological offices.

Although ADS itself does not specifically encompass ATS communications, automation or procedures, clearly all of these elements must be tailored to support the ADS function and to make meaningful use of the data. For this reason it is useful to consider the ATM automation and communications systems as the foundation upon which an ATC system, using ADS, will be built.

ADS will significantly change the way ATC is performed in oceanic and other areas which are beyond the coverage of land based radar, line-of-sight communications systems or combinations of these.

The accurate and timely indication of an aircraft's position and good communications are keys to the operation of a safe, responsive and effective ATM system. Given the capabilities of ADS, the air traffic controller will be better able to monitor flight progress, ensure safe separation of aircraft and respond in a timely manner to user requests. A gradual transition from procedurally oriented ATC, where such is being carried out, towards a more tactical control environment will also be possible with the introduction of ADS. Figure 5.1 displays an overview of ADS.

Pilot interface

The ADS data reports are transmitted automatically without pilot action. The frequency of reporting is determined by the ATM system, however, a capability will be provided to permit emergency messages to be initiated by the pilot by means of a simple operation using quick action facilities. Additionally, pilot interface will provide a means to send and receive routine data messages. Direct pilot-controller voice capability must be available for emergency and non-routine messages.

Avionics

The ADS function is supported by avionics equipment which is able to gather aircraft data from on-board systems, format them and direct them to the relevant air-ground link via the data link processor. The avionics will make maximum use of equipment already in place on most commercial aircraft.

On-board equipment will also have the capability to receive messages originated by the controlling ATM unit which will set the position report update rate of an aircraft and the data fields to be included in the report.

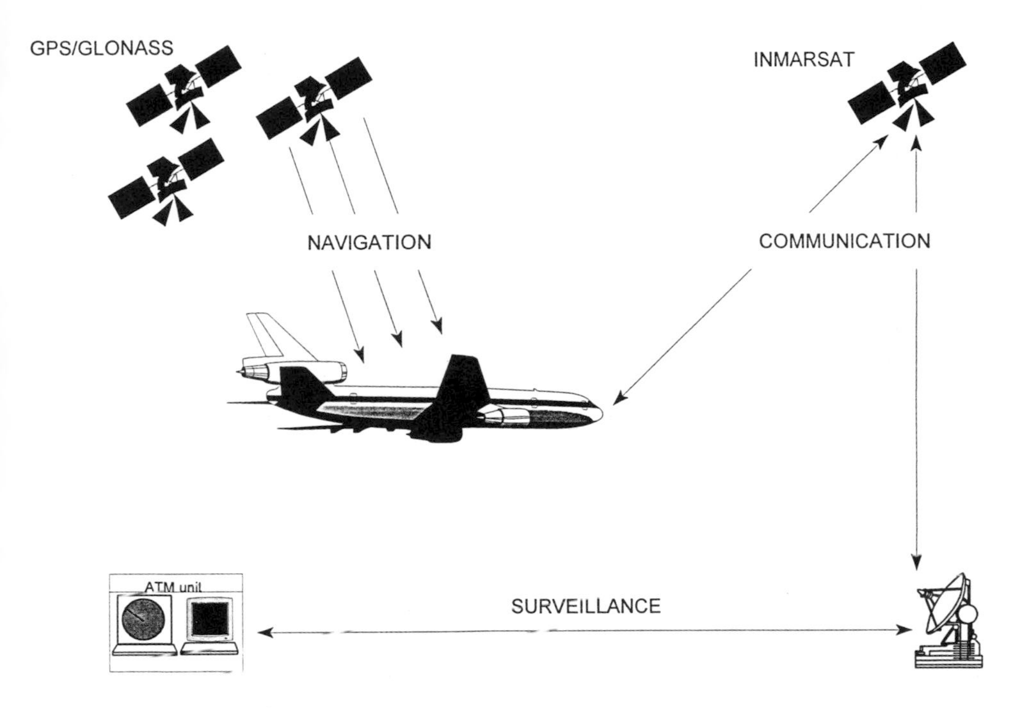

Figure 5.1 ADS overview

Data links

A complete end-to-end communications sub-system operating between the ATM unit and the aircraft is presently being designed to effectively support ADS requirements. The architecture of such a communications sub-system will permit a modular design and implementation and will accommodate an incremental system development approach. At the same time, functional uniformity in the services provided to airspace users can be maintained while allowing individual ATS providers considerable flexibility in the actual implementation.

The FANS Committee recognized that a satellite-related ADS service could be provided using any of the following four modes:

* Contract mode, in which the aircraft automatically provides ADS reports at prearranged intervals or upon reaching preset navigational limits;

* On-demand immediate response;

* Time-ordered polling; and,

* Random access.

Contract mode is expected to be the mode used for ATM purposes. It will permit the ATM system to instruct aircraft to report their positions periodically. Aircraft may also report at any time or be interrogated by the ATM system when so required by operational circumstances.

Communications interface

The efficient implementation of an ATC system in an ADS environment requires a ground communications interface connecting ATM units with the associated data link ground station(s). One of the functions of this interface will be to route the ADS messages to the appropriate end users (ATM units). Direct routing to multiple end users will be accomplished by this interface and will not require intermediate processing by the ATM unit. However, the logic for this routing will be controlled by the appropriate ATM authorities. The communications interface should permit ATC systems where ADS is available to accept data via any of the data link media: satellite, VHF, HF and Mode S (see Chapter 3 for a description of data link and the associated aeronautical telecommunications network (ATN)). Figure 5.2 displays the ADS architecture within the ATN.

ATM automation and ADS

An ATM data processing system where ADS is available should be capable of automating the following functions:

* Flight data validation, whereby the aircraft's route (waypoints) entered into the aircraft's navigation system is compared to the cleared flight profile and any discrepancies found are reported;

* Conformance monitoring, whereby the aircraft's reported actual and intended positions are compared to the cleared flight profile and out-of-tolerance conditions reported to the controller;

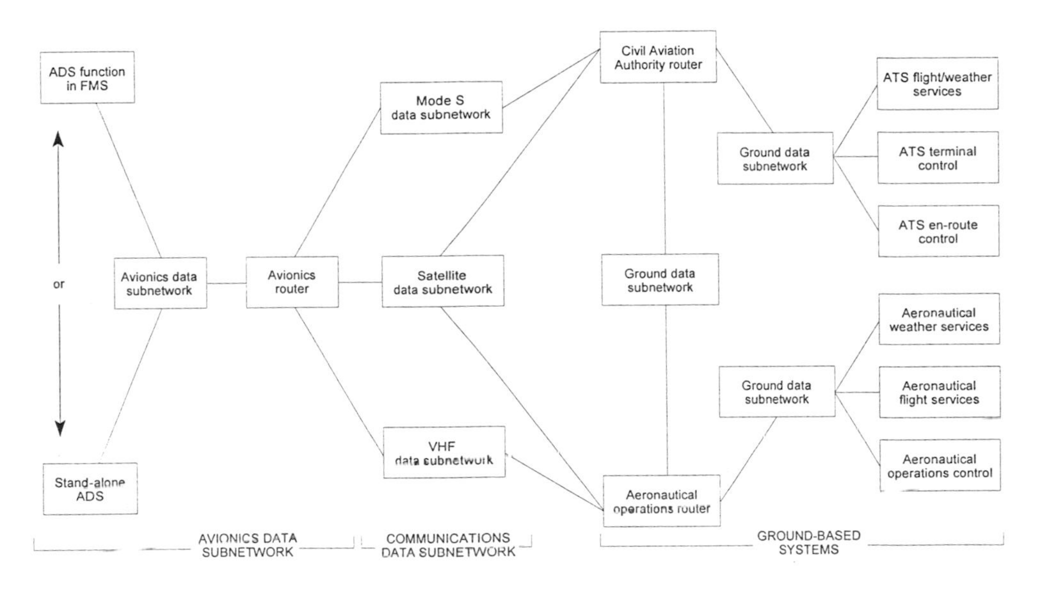

Figure 5.2 ADS architecture within the ATN

* Automated tracking, in which ADS-reporting aircraft are tracked by the ATM system between position reports;

* Detection of potential conflicts in which positions of all aircraft are projected into the future and a search is performed to detect possible future violations of separation minima;

* Conflict resolution, whereby a system solution to a potential conflict is offered to the controller which can either be used of modified; and,

* Presentation of relevant processed data to the controller.

These functions do not necessarily represent a complete set. It should be noted that some of these functions may need to be performed by a controller depending on the level of automation of the ATM system.

Controller interface

The human-machine interface will be left to the individual service provider, however, ICAO has developed a human factors programme that is addressing issues related to CNS/ATM from many perspectives (Chapter 8 addresses human factors issues in more detail). Briefly, all such interfaces should be designed and integrated into existing ATC systems, with the objective of reducing controller workload. The interfaces should have the following capabilities:

* To display the traffic situation so as to enable a controller to monitor the traffic in the sector with minimum effort;

* To alert the controller to potential conflicts;

* To enable the controller, using both predetermined and free format, to compose and transmit ATM messages via data link;

* To display the arrival and content of data messages to the pilot;

* To provide rapid access to a voice channel for emergency and non-routine communications with a pilot or a group of aircraft; and,

* To provide a method of responding rapidly to a request from a pilot via voice communications.

Future development

The concept of ADS is not limited to the oceanic environment. ADS can be used in other areas such as over deserts, jungles, mountainous regions and, in fact, anywhere there is satellite or VHF data link communications coverage. This in itself is economically advantageous, since satellite-based communications and surveillance do not require as large an initial capital investment as radar and VHF communications-based systems. Operations and maintenance costs are also expected to be lower.

Since ADS transmits position information derived from an aircraft's on-board navigation system, independent of the type of navigation system employed, highly accurate satellite navigation systems (i.e., GPS and GLONASS), can be integrated into an ATC system where ADS is available.

The use of GPS and GLONASS, in conjunction with ADS, will support a reduction in separation minima, since the effect of each individual aircraft's navigation system performance and deviations will be minimized.

ADS Broadcast (ADS-B)

ADS-B is an application of ADS technology which resulted from the extensive research and trials that have been carried out over the past few years. Using ADS-B, an aircraft suitably equipped, broadcasts its position, altitude and vector information for display by other aircraft and also by ground users, such as ATS providers. Under the concept, each ADS-B capable aircraft or ground vehicle would periodically broadcast its position and other required data available from on-board equipment. Any user, either airborne or ground-based, within range of this broadcast, could receive and process the information. The aircraft or vehicle originating the broadcast would have no knowledge of what or who is receiving its broadcast. Everyone in the system would have real-time access to precisely the same data, via similar displays, allowing a vast improvement in situational awareness (Daly, 1996).

ADS-B is a source of positional information and is not the same as the surveillance applications described earlier in the chapter. Other services may be provided to aircraft via broadcast such as delivery of flight information services.

Because it is a broadcast system, ADS-B requires a data link to support its performance requirements. Several possibilities are being explored, such as using the Mode S data link capabilities or available VHF channels.

In a fully equipped environment benefits could be envisioned as follows:

* Enhanced situational awareness where users with a traffic display would have the ability to visually acquire aircraft within range allowing a more strategic knowledge of the traffic situation, up until now available only to air traffic control;

* Aircraft and ground-based conflict detection and resolution, supporting future ATM systems where pilots and/or aircraft systems would perform conflict detection and resolution functions;

* Reduced infrastructure costs where radar equipment could be dismantled as the ADS-B would serve the surveillance function. Additionally, radar separation services could be extended to airspaces not presently covered by radar.

By allowing a greater degree of situational awareness in the cockpit, the aircraft could become more involved in ATM. Using ADS-B in this way an aircraft could specifically identify other aircraft, such as a leading aircraft, along with other relevant information. Delegation of responsibility for separation could theoretically be transferred from the ground to the cockpit. Additionally, in combination with computer software, ADS-B information could support conflict detection and resolution both in the air and on the ground. This would include conflict detection on the airport surface, for example, in predicting potential runway incursions.

For Visual Flight Rules (VFR) pilots, the ADS-B information could be used as an aid to situational awareness and visual acquisition and support a type of "electronic VFR" (RTCA, 1996).

ADS-B has the potential to significantly transform ATM by allowing substantial efficiency gains in high density airspace.

Much work remains to be done regarding ADS-B, both conceptually and technologically. As the technology matures, an operational concept and associated procedures will be developed for integration into the operational environment. This will be a gradual process and eventually ADS-B will form part of an ATM system, described in Chapter 6, where flight operations are fully integrated.

Still to be resolved is the issue of what radios and frequencies should be used for the ADS-B data link. A leading contender is a technique that could be easily introduced using the Mode S transponder from equipment that is already installed in most commercial aircraft to support the Traffic Collision Avoidance System (TCAS). Other methods would use a variety of time-division multiple-access (TDMA) techniques, one using Mode S, and the self-organizing TDMA over VHF data link, which is discussed later in the chapter under the section: A glance at the future.

Secondary surveillance radar (SSR) Mode S

> SSR is a well proven technology, which has been in use in one form or another in civil aviation operations for about thirty years (ICAO, 1993a, p. 4-17).

Radar in general, and particularly SSR, is in wide use in many parts of the world where terrestrial surveillance systems are appropriate, which includes terminal airspace and high density continental airspace. With CNS/ATM systems, traditional SSR as well as SSR Mode S, will be used for the foreseeable future.

Mode S is an enhanced mode of SSR that permits the selective interrogation of Mode S transponders and the two-way exchange of digital data between Mode S ground stations and transponders, and also the interrogation of other transponders.

SSR Mode S offers two distinct facilities. The first is an improved SSR surveillance capability through discrete addressing (each aircraft equipped with a Mode S transponder operating in a Mode S environment will have its own unique mode, or address). In this same category, Mode S has the ability to augment surveillance information with related data, such as aircraft intention information, which assists radar tracking (ICAO, 1993a). Secondly, SSR Mode S offers a general purpose ATS air-ground data link capability.

Surveillance enhancement

The SSR Improvements and Collision Avoidance Systems Panel (SICASP) (ICAO, 1988c) describes the enhancements to be expected by incorporating Mode S data link into radar data processing systems as follows:

Aircraft identification Operational identification of aircraft is currently performed by the assignment of reusable discrete Mode A codes. the number of different codes provided by the existing SSR system is limited to 4,096. Two disadvantages result from this method of identification:

* A separate correlation must be made between the Mode A code associated with the surveillance plot and the call sign as indicated in the flight plan; and,

* Assignment of unambiguous Mode A codes is limited to those flights to which radio communication has been established.

Furthermore, there are areas of the world where more than the available number of different Mode A codes are required to cope with operational needs.

With Mode S data link it is possible to report directly, the aircraft identification as used in the flight plan. When no flight plan is filed, the aircraft registration is reported. Any ambiguity as to the radar identification of an aircraft through a Mode A identity code is thus removed. The reported aircraft identification can be either:

* Fixed, for aircraft that normally use their registration as the call sign; or,
* Set by the pilot for aircraft that normally use a flight number.

Aircraft altitude The altitude code structure employed in Mode S surveillance replies provides the capability for reporting altitude in 25 ft. increments. Furthermore, because of the error protection employed, only one valid reply is required for an altitude report. When aircraft possess suitable sources of digital data, the link can be used to provide an alternative altitude report on demand for confirmation purposes.

Aircraft status Mode S data link enables aircraft to report either on the ground or in the air. Such information can be useful for radar data processing, automatic association of flight plan and radar data, as well as airborne collision avoidance system (ACAS) surveillance processing.

Aircraft trajectory parameters In some ATC systems, radar trackers are used to predict the future position of aircraft for the purpose of automatic conflict detection and other functions. The performance of these functions can be enhanced by the downlinking of such information as heading and roll angle which allows early detection of aircraft manoeuvres.

Finally, by enhancing SSR with Mode S, the selective address and data link capabilities will further improve the beneficial role of SSR for surveillance purposes. Figure 5.3 depicts a complete ADS based ATC system.

Summary of the benefits of the future global surveillance system

* ADS service will be the basis for potentially significant enhancements to flight safety;

* ADS will provide significant early benefits in oceanic and other non-radar areas where voice position reporting is the only available means of surveillance. Use of ADS, supported by direct pilot-controller communications, will allow these non-radar areas to evolve to the point where ATS is provided in the same manner as in today's radar controlled airspace;

* ADS will support reductions in separation minima in non-radar airspace. These reductions will alleviate delays and minimize necessary diversions from preferred flight paths thereby reducing flight operating costs;

* ADS will support increased ATM flexibility, enabling controllers to be more responsive to aircraft flight preferences. With or without reductions in separation minima, this flexibility will contribute to cost savings for flight operations;

* Mode S will provide high accuracy, reliable surveillance in high density airspace;

* SSR in combination with ADS will facilitate uniform surveillance service worldwide (ICAO, 1993a). Figure 5.4 shows the differences between the present and future surveillance systems.

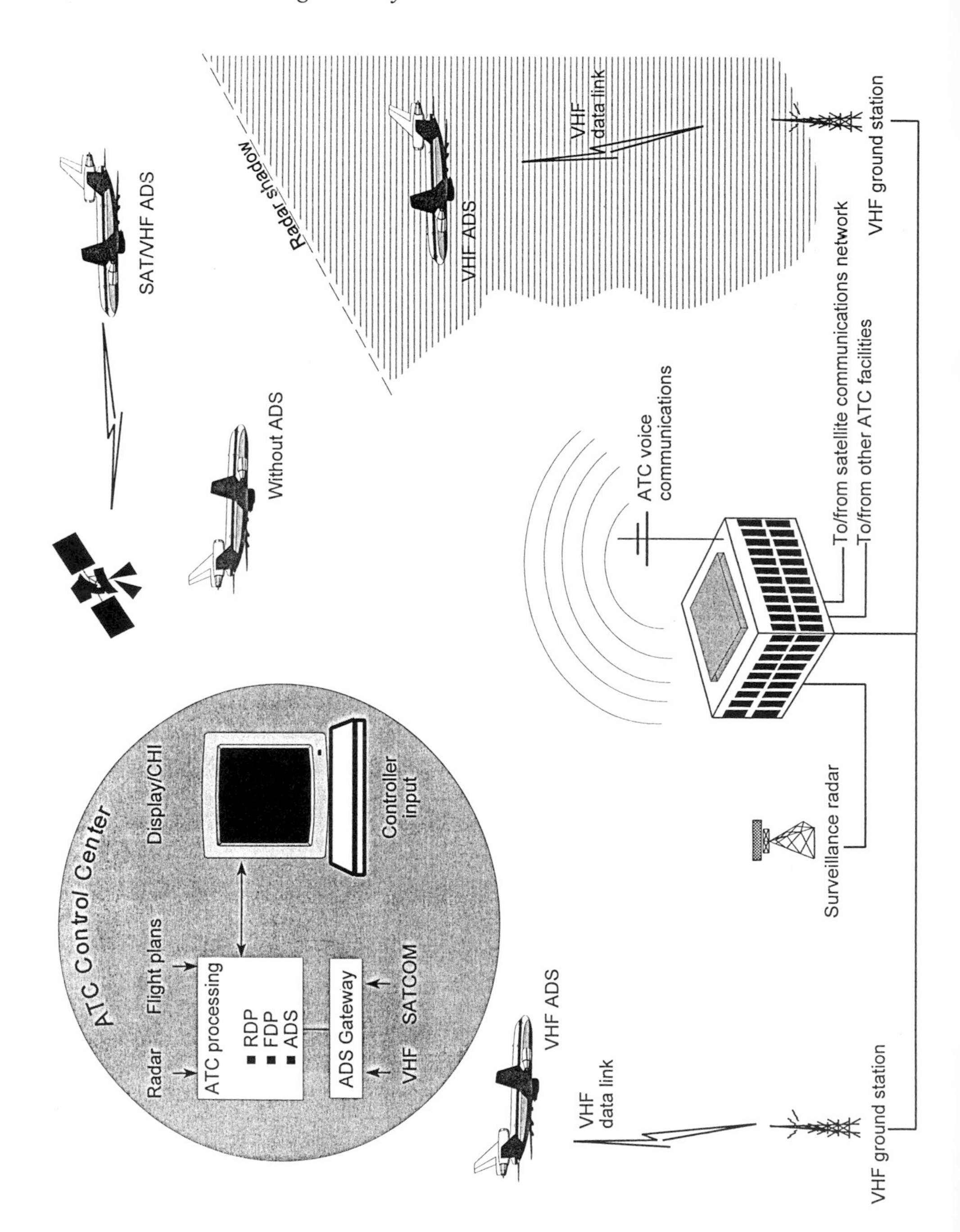

Figure 5.3 An ADS based ATC system

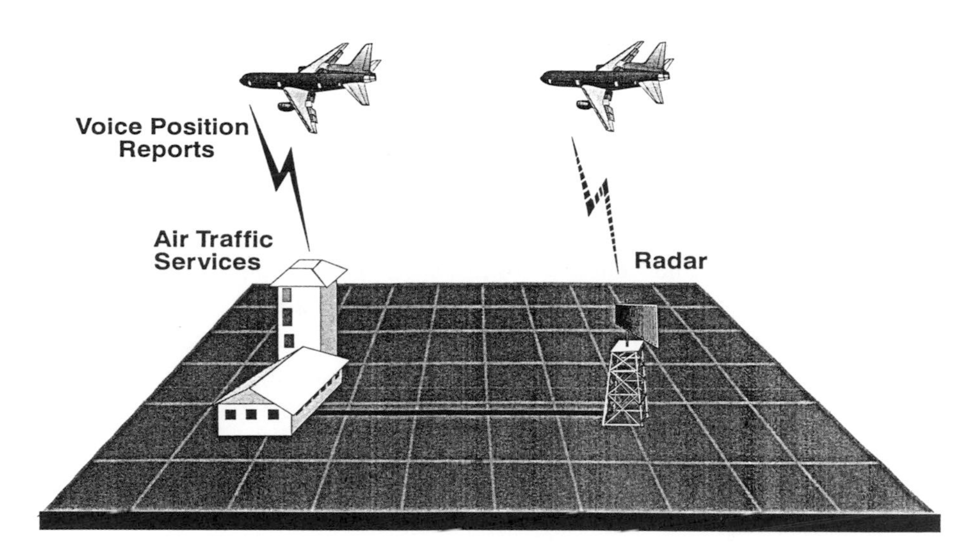

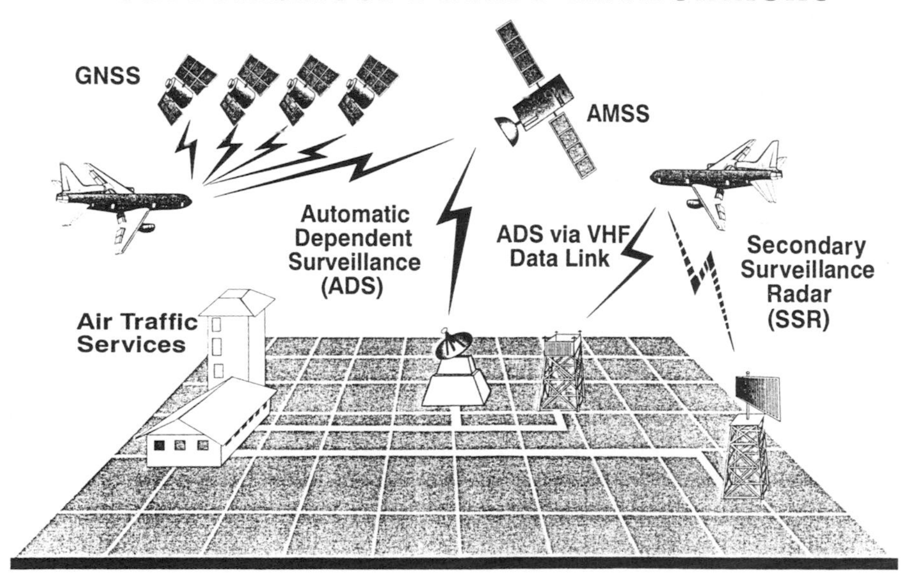

Figure 5.4 Differences between the present and future surveillance systems

A glance at the future

> ADS-based ATC systems will provide controllers with the tools to be more responsive to users (Castro-Rodriguez, 1990).

Over the oceans

Chapter 3 went into some detail concerning the importance of ICAO's regional planning bodies and in particular, their planning and implementation activities concerning the communications element of CNS/ATM systems for the North Atlantic and Pacific regions. This chapter stressed the relationship between the elements of communication, navigation and surveillance. This relationship is particularly important concerning communications and surveillance as both Mode S radar and ADS have both communication and surveillance functions and will serve both purposes in an integrated manner.

Given the difficulties associated with both communications and surveillance over the oceans, it is no surprise that these regions are also leading the way in experimentation and trials of ADS. Chapters 3 and 4 described these activities in some detail.

Dynamic oceanic track system (DOTS) and Dynamic aircraft route planning system (DARPS) in the Pacific Region The limitations of oceanic ATC communications and surveillance currently confine trans-Pacific aircraft to tracks which are revised only once every twenty four hours. ATC deficiencies are based mainly on the fact that procedures still require manual entry of aircraft position based on pilot position reports. It therefore falls to the controller, with no automatic backup, to ensure that aircraft conform to their flight plans. Under normal circumstances in oceanic regions, the controller accomplishes this task with no visual display of aircraft position and/or intentions, but must visualize the entire situation, and the position of all aircraft, from aircraft position, time and altitude reports.

In an attempt at improving the situation, the airlines operating in this region worked closely with the ATC provider States through several programmes aimed at the introduction of dynamic rerouting of aircraft in the Pacific Region.

The idea is based on the premise that crews of aircraft, equipped with the latest weather data, are able to calculate and request more advantageous tracks and flight levels. It is the near real time knowledge of aircraft position delivered by ADS that will allow the controller to give the required clearance (ASA, 1992).

Through its own dynamic aircraft route planning system (DARPS) programme, United has shown that it was feasible to replan a flight after aircraft departure. DARPS was designed to provide the crew with the best available flight plan and to ensure the constant monitoring of flight progress from departure to arrival. With full implementation of DARPS, real-time monitoring of aircraft performance, combined with the ability to replan and modify the route, makes it possible to continuously recalculate actual arrival time and fuel remaining at touchdown. This will permit reductions in contingency fuel leading in turn to improved payload capability.

The central premise behind DARPS is ADS. ADS will have the effect of changing the function of the air traffic controller from applying strategic airspace restrictions to taking responsibility for the tactical management of aircraft and airspace. As described in Chapter 3, the States of the Informal South Pacific ATS Coordinating Group (ISPACG) developed the FANS 1 implementation plan (Price, 1994) with identified benefits and goal dates for implementation of flexible tracks, dynamic rerouting capabilities and reduced lateral and longitudinal separation.

As the initial implementation of ADS in the Pacific will be by means of the FANS 1 package, and the FANS 1 package integrates the ATC data link into the flight management system (FMS), ADS messages will not have to be laboriously assembled from several different sources. For example, the primary source of position and time data, GPS, is itself integrated into the FMS, thereby allowing for the direct transmission of this information from the FMS to the ground system.

The North Atlantic Region experimentation and trials The North Atlantic trials began in 1993 as the result of an agreement between the United States and the United Kingdom to conduct trials of ADS and data link communications over the Atlantic Ocean. Results of the North Atlantic trials will help validate requirements and procedures such as those outlined in the ICAO Standards and Recommended Practices (SARPs).

As part of the North Atlantic Systems Planning Group (NAT SPG) activities, several States in that region have been experimenting with ADS as part of an ongoing effort to assess the usefulness of ADS over the North Atlantic (Price, 1994). Unlike the Pacific, the North Atlantic is a smaller region with less airspace and more condensed traffic flows between two highly developed and crowded regions on either side, Europe and North

America. As the overall airspace is considerably smaller, the cost savings from dynamic reroutes in the North Atlantic would not be as beneficial in the near term as in the Pacific Region. Greater savings and efficiencies will more likely come however, from decreased separations, realization of which is being aggressively pursued in that region as is evident with plans to reduce the vertical separation minima above flight level 290 by 1997.

It is therefore quite possible that States serving the North Atlantic, along with the airlines that operate in this region, will forego a decision to implement CNS based on the aircraft communications addressing and reporting system (ACARS) network described in Chapter 3. It is more likely that the North Atlantic Region will adopt a philosophy that encourages speedy development and implementation of the ATN, also discussed in detail in Chapter 3, thereby taking full advantage of a thoroughly integrated CNS system. The States in the North Atlantic Region are however, continuing a programme of ADS experimentation and trial. The United Kingdom Civil Aviation Authority has been investigating the application of satellite communication and ADS trials (ICAO, 1993a) using a British Airways 747-400 suitably equipped with avionics equipment, flight management computer (FMC), GPS sensor units, a satellite data unit, high gain antennas and a computer platform to perform ADS and routing functions aboard the aircraft.

Ireland has been conducting engineering trials on the integration of High Frequency (HF) reports, radar data and position reports delivered by ACARS (ICAO, 1994f) and on presenting the information in graphical form on a work station.

Canada has been involved in extensive ADS experimentation and trials in the North Atlantic as well as over its northern airspace for several years using VHF, satellite and HF data link. One element of this work involves the development of a prototype work station to meet the needs of the oceanic air traffic controller in the future system (MacLean, 1994). This system has been designed to complement existing oceanic flight planning systems and to support ADS and will provide the controller with a traffic situation display similar to radar. The display will show the estimated position of an aircraft based on the last position report and the flight plan route. For ADS reporting aircraft, the estimated position is based on the last ADS reported position. Additionally, specific software will allow projection of the flight information into the future at a faster than normal rate to provide the controller with advanced knowledge of the developing traffic pattern.

Sweden

At the First Meeting of the GNSS Panel, Westermark (1994) described the Swedish experimental system. In this system, navigation, communications and surveillance elements are being amalgamated into one unit in order to realize the benefits of an integrated ATM system. This integrated system is now known as the *Cellular Concept* (ICAO, 1994g). The complete system has become known as the GNSS-synchronized, self-organizing, time-division, multiple-access (STDMA) data link.

The key component of this concept is the *GNSS transponder*. This unit is installed in mobile stations (e.g., aircraft and airport vehicles) and ground base stations. The GNSS transponder is comprised of a commercial GPS receiver, a VHF transceiver and a communication processor. The two latter components with associated software make up a data link. Interfaced with a satellite transceiver, the GNSS transponder would be further capable of supporting AMSS applications, such as ADS reporting in areas outside VHF ground coverage.

The GNSS receiver serves multiple purposes in the GNSS transponder architecture. The position information derived by this unit is not only used for navigation and to report position to the base station in the experiment, but also as an element in the communications logic. Moreover, the accurate satellite time (each GPS satellite carries an atomic clock) which is output by the GNSS receiver, is used by the communication processor to synchronize all users to a global time base. The use of the accurate satellite time is a unique feature of the concept being developed by Sweden and is essential for achieving the capacity and robustness needed in time critical applications.

This experimental data link concept utilizes time division multiple access (TDMA) technology, which means that communications on the data link are synchronized in time and that each message is allocated a defined time slot. This slot allocation is based on accurate time obtained from the GPS satellites and enables users to transmit their messages in very precisely defined time slots and causes a reduction in the gaps between slots (Daly, 1996).

A satellite receiver is available in the experimental base stations and serves as a differential GNSS (DGNSS) to compute differential corrections. The differential corrections are transmitted to the mobile stations (e.g., aircraft, airport ground vehicles) via the data link. Differential corrections uplinked every second can be easily accommodated on the same VHF channel as that used for broadcast of position information.

The integrity of the navigation solution in mobile stations is ensured by receiver autonomous integrity monitoring (RAIM) facilities.

Installed in aircraft and ground vehicles, the GNSS transponder determines its position based on the GPS satellites adjusted by the differential corrections transmitted from the base station. The position information is then transmitted on the data link to other aircraft and ground vehicles operating in the system, and to the base station. Specially installed cockpit equipment can then allow air crews to monitor the position of other aircraft and vehicles. This is in support of ADS-B described earlier in this chapter where the positions and intentions of all aircraft in the air and on the ground are constantly broadcast, by datalink, to each other and to air traffic control.

This system forms the basis of an ADS system, allowing controllers to view the traffic situation as they would on a radar display, combined, if desirable, with SSR data. As GNSS transponders can also be fitted in airport vehicles, the system forms the basis for an advanced surface movement guidance and control system (ICAO, 1994g).

STDMA is being evaluated as part of a project known as the North European ADS-B Network (NEAN). The work programme is to demonstrate the benefits of ADS-B systems to the users, to develop and validate the technology for ATM and to establish a cost database. Eventually, as many as fifteen aircraft are expected to be involved. Ground stations are located throughout northern Europe.

Other experimentation

Agency for the Security of Aerial Navigation in Africa and Madagascar (ASECNA) ASECNA has been involved in activities on ADS related operational and technical trials within the context of implementation of an integrated CNS/ATM system. Several Air Afrique Airbus A310 aircraft have been equipped with satellite communications equipment. ADS messages are transmitted using the ACARS VHF and satellite communications. A controller interface has been chosen and the trials will be replaced by the preoperational ADS for Africa system known as AFIADS.

Australia In addition to its work in the Pacific Region, Australia will commission an ATC system in 1997 which will incorporate both ADS and controller pilot data link communication (CPDLC) (Challinor, 1994). It is expected that ADS will be very useful to Australia over its vast, inhospitable inland regions where the establishment and maintenance of radar equipment is not feasible.

Australia's ADS and data link interim system (ADIS) project, implemented in the Pacific Ocean sector in mid-1996, provides ATS using digitally encoded communications, via satellite or VHF data link, with aircraft equipped to receive the signals. This will allow aircraft to use more fuel efficient routes and allow a greater use of flexible tracks (Castro-Rodriguez, 1994).

China The research, development and trials for ADS in China are geared toward the development of suitable software to support ADS message processing and also to support ADS system trials and implementation. The goal is to develop an air navigation processing system using VHF data link, satellite data link, a ground earth station, the ground communication network and a computer processing system, to process ADS messages in adequate time and to display the results to users (Castro-Rodriguez, 1994). The Civil Aviation Administration of China (CAAC) has integrated CNS/ATM into its northern region facility at Beijing as part of a program that includes Boeing, United FANS 1 B747-400s, an Air China GPS equipped B737 and a workstation provided by Raytheon. Data networks provided by ARINC and SITA are used for the provision of these services. Initial ADS trials in the spring of 1996 were followed by ADS/radar integration in June. An upgraded system is scheduled to begin flight trials later in the year. Following the trials, the CAAC intends to implement the first operational CNS/ATM ATS route along the Chinese-Russian border.

France The French trial of ADS began in the summer of 1993 with controller familiarization activities. The prime objective of the trial is to examine the transition from ADS to radar airspace on the approaches within the Brest ATC centre in France along the Atlantic coast (ASA, 1993a). Additionally, France has installed a work station in order to assist in its evaluation of the integration of ADS and radar data and has also begun simulations to study various ADS related issues (ICAO, 1994f).

Iceland continues with its VHF and HF data link trials in oceanic airspace which includes ADS trials concerning the transmission of binary data over a very high frequency (VHF) data link and the integrated display of ADS and radar data. Based on the successful results of the trials, Iceland believes that HF data link could be a valid medium for ATS applications.

India has made a major commitment to CNS/ATM. Trials have been conducted in Madras using data link position reports from specific Qantas aircraft and development is ongoing with VHF data link, avionics and communications. India has contracted the Raytheon company to modernize its systems in the Delhi and Bombay air traffic control centres which address the needs for ATM using data link for communications and also implementation of Mode S with data capability. Perhaps the most significant step taken was the provision of an upgrade to their new ATC automation equipment which will allow FANS 1 equipped aircraft to make use of that equipment while India begins demonstration of an operational concept for ATC based on the updated equipment (Arkind & Medis, 1996).

The Indian trial forms part of the CNS/ATM regional implementation plan for the Asia-Pacific, part of which calls for ADS demonstrations in India. It is based on a memorandum of understanding concluded with the Australian Civil Aviation Authority (ACAA) which sponsored development of some of the systems to be used. ACAA has similar agreements with Indonesia and Singapore. "This means that we will soon have ADS from Australia to Singapore", says ACAA general manager of research and development, Brian O'Keefe (ASA, 1994b, p. 7).

In addition to the above, and as a result of heavy traffic over India's Bay of Bengal region, India is scheduled to implement an interim ATM system for this area. Although this interim system is designed to handle aircraft equipped with the Boeing FANS 1 package, it has the added feature of being able to display non-ADS equipped aircraft using the voice position reports, thereby providing controllers with the ability to display all aircraft using Bay of Bengal airspace. This equipment is expected to lead to a reduction in departure delays and enable the creation of a new integral air route over the region, which will allow aircraft to avoid the crowded airspace over Calcutta.

Italy launched its ADS trial in the Mediterranean area in October 1992 in order to demonstrate the technologies needed to fill the ATC surveillance gaps in the Mediterranean area. The objectives of the programme are reduced separations and thus, increased traffic and a more efficient use of the airspace (ASA, 1993b).

The main purpose of the Italian ADS programme is to develop an ADS/ATC system to be introduced in the Mediterranean area (Tomasello, 1994). The main objectives are towards:

* The development of a common controller work station, to be used regardless of the data link sub-networks (i.e., AMSS, Mode S or VHF data link);

* The development of associated software (e.g., ADS tracking, human machine interface for data link and data link ATM applications); and,

* The connection of the work station into the evolving ATN ground network while also permitting integration into the presently configured area control centres.

It is well known that the majority of the aircraft operating in the Mediterranean area will be equipped with a VHF airborne radio, while most routes in the area are completely within VHF coverage. It is therefore expected that VHF data link will support the ATM function in this area.

Japan The Japanese ADS satellite data link experimental programme aims at the evaluation of test results of ADS transmission performance, including ADS-derived aircraft position and transmission delays. Three major airlines of Japan serving international routes and the Japanese Civil Aviation Bureau fully support the trials and the early introduction of CNS/ATM with the aim of making the best use of available technology (Castro-Rodriguez, 1994).

Norway has carried out trials involving a satellite based ADS system for monitoring helicopters operating over the North Sea. The objective of the Norwegian trials is to develop ADS equipment that is tailor made for use on helicopters. If the trials are successful, the equipment will be made mandatory on all helicopters operating in the Norwegian sector of the North Sea.

North Korea Another obstacle to more direct routes in the Asia/Pacific region is slowly being overcome. Through the efforts of ICAO and the International Air Transport Association (IATA), North Korea has recently (mid-1996) signed an agreement that will permit flights to operate through its airspace beginning in December 1996. As part of the agreement, North Korea will improve communications support for FANS 1 equipped aircraft (Nordwall, 1996).

Russian Federation The Russian Federation plans on using ADS as the principal means of surveillance over the vast regions of the Russian North, Siberia and the Far East. By the year 2000, the Russian Federation is expected to have established four experimental CNS/ATM ATC centres using all available CNS/ATM technologies. New technologies will be integrated as they are developed.

Through the use of ADS, the Russian Federation expects to be able to greatly reduce the number of existing ATC centres (ICAO, 1995d).

On September 2, 1995, as part of a series of ongoing trials and demonstrations, a United Airlines flight overflew the Russian Far East, enroute from Chicago to Tokyo, while conducting controller/pilot data link communications (CPDLC) and ADS trials with one of the Russian experimental CNS/ATM ATC centres located in Magadan, Russian Federation. The United flight was the first ever to fly in the Russian airspace using the FANS 1 capabilities of the Boeing 747-400 and the first to use FANS 1 on a non-oceanic route (CNS Outlook, 1995). The aircraft and equipment used were developed for use in the Pacific Region as described in Chapter 3.

The FANS 1 package enabled the United flight to automatically report its position information via ADS, in effect, allowing the air traffic controllers in Russia to see the flight on a radar-like display from immediately after departure at Chicago, to touchdown in Japan.

Future flights are expected to take place deeper into Russian Federation territory. United expected to fly five such flights a day by the end of September 1995.

The Russians are so sure of the possibilities afforded by ADS that they have stated publicly they will not make major investments in enroute radar systems, but will rely on ADS and CPDLC for tracking and controlling aircraft traversing their airspace. Russian Federation President Boris Yeltsin has reportedly issued a government decree to accelerate the opening up of Russian airspace to foreign overflights and to affirm that all new routes will be based on CNS and not radar (CNS Outlook, 1995).

Singapore To meet the challenges of the twenty first century, the civil aviation authority of Singapore has invested approximately 120 million U.S.

dollars on equipment and buildings for a new air traffic control system known as the (second generation) long range radar and display system (LORADS II) designed and installed by Thompson CSF. This system is in the final phase of implementation and is designed to exploit the technology associated with CNS/ATM systems (Eng, 1993) and is designed in a manner to accelerate Singapore's transition to the global air navigation system as developed by ICAO.

The LORADS II processes and presents ADS data received from aircraft via satellite. The ADS reports will be validated against the cleared flight profiles stored in the system and presented to controllers. The ADS software will have a number of monitoring functions, such as conflict prediction, warnings whenever aircraft intentions are inconsistent with the cleared flight profile, and warnings of missing ADS reports. The system also supports two-way data communications between pilots and controllers and controllers will be able to transmit and receive operational messages between ATS facilities via satellite data link.

Spain The idea of integrating SSR and ADS data with the purpose of enhancing the surveillance function, increasing the level of surveillance availability and monitoring the integrity of the navigation system, was presented by Spain to the FANS Committee (Diez 1993a). Spain has in fact carried out a study with the objective of developing an adaptable tracking algorithm which would process both SSR and ADS data.

It has been determined that SSR data, when combined with ADS data, can upgrade the performance of the surveillance function to a level similar to that of SSR Mode S (Diez 1993b), in addition to providing coverage at low altitudes and other blind areas. It also provides flexible redundancy, ensuring that the system reaches an adequate level of availability in a cost-effective manner, and providing the means to monitor in a timely way, the integrity of the navigation system being used by the aircraft.

Additionally

- Demonstrations of ADS are underway in several other States and regions not mentioned above;

- The ADSP is responsible for development of the technical and operational requirements for ADS (Castro-Rodriguez, 1994), with a view toward developing SARPs, procedures and guidance material on ADS and to provide timely information exchange with other ICAO bodies (ICAO, 1995c). SARPs for ADS based procedures are being developed for completion in 1996 and 1997;

* The ADS Panel is developing an ICAO manual for ADS and ATS data link applications which will define the related operational requirements, including guidance for States implementing CNS/ATM systems. This will include material on automated terminal information service (ATIS) and predeparture clearance;

* The ADS Panel is also developing the operational concept for ADS-Broadcast, which will provide robust and low-cost line-of-sight surveillance via VHF or Mode S broadcasting of position information from suitably equipped aircraft and ground vehicles to ATC and to each other;

* Operational use of SSR Mode S is already beginning in the ICAO North American Region, will begin in the Pacific Region in the late 1990s and in the European Region by the year 1999;

* The AMCP has developed SARPs for the air-ground communications data link and its sub-networks (e.g., AMSS, VHF and SSR Mode S). Additionally, the potential of HF data link to support ADS and ATS applications is being investigated;

* The RGCSP is working toward development of a methodology to determine separation standards for an ADS-ATC system. It is probable that ADS will provide the means for reduced separation standards in certain types of airspace (Malescot and Chenevier, 1993).

Recapping the major points

Previous chapters have shown how aircraft operating in a CNS/ATM environment will communicate with people and systems on the ground and navigate. For effective ATM to be possible however, people or systems on the ground must know the position of aircraft on a continuous basis and be able to estimate their future position.

Generally, the idea of keeping track of an aircraft's position is referred to as surveillance. With the introduction of air ground data links together with accurate and reliable aircraft navigation systems, a new method of surveillance, known as ADS has evolved and will form an integral part of CNS/ATM systems.

This chapter described ADS and how its integration with the communication and navigation components of CNS/ATM systems, described in previous chapters, will lead to new and innovative surveillance capabilities.

A review and analysis of Mode S radar, also forming an integral part of the CNS/ATM surveillance element was also made.

Finally

* SSR Mode A/C will continue to be used in terminal and in high density continental airspace as SSR is presently in wide use in many parts of the world where terrestrial systems are appropriate;

* SSR Mode S, with its selective address and data link capabilities will further enhance the beneficial role of SSR for surveillance purposes;

* The introduction of air-ground data links, together with sufficiently accurate and reliable aircraft navigation systems, will present the opportunity to provide surveillance services in areas which lack these services in the present infrastructure. This surveillance capability is known as ADS;

* ADS will be used mainly in non-radar coverage areas, however, it will also find beneficial application in other areas, including high density areas, where ADS may serve as an adjunct and/or backup for SSR and thereby reduce the need for primary radar;

* Although the use of primary radar is declining, its utility will continue in those airspaces where there is a mix of SSR equipped aircraft with non SSR equipped aircraft. However, the overall need for primary radar will diminish with the gradual introduction of SSR Mode S and ADS.

Questions and exercises to expand your knowledge

1) You should now have an understanding of the relationship between the various components of communication, navigation and surveillance as envisioned in CNS/ATM systems. Describe how the communication and navigation elements can be integrated to permit ADS to be put to effective use by both the ATS providers and the aircraft operators. In your answer you should include at least the following terms:

    * GNSS;
    * Data link;
    * Satellites;
    * HF;
    * VHF;
    * Mode S radar;
    * ATN;
    * FMS.

2) What advantages would ADS offer in low density airspace and how would ADS offer those advantages?

3) What advantages would ADS offer in high density airspace?

4) Do you think it is cost-beneficial to make ADS available over all the oceans, deserts and other remote parts of the earth? Justify and expand upon your answer.

5) Briefly describe the scenario as envisaged by the FANS Committee regarding surveillance in future CNS/ATM systems.

6) Describe in your own words the difference between independent surveillance and dependent surveillance. Are the following types of surveillance considered as being dependent or independent:

    * PSR;
    * ADS;
    * SSR;
    * Mode S SSR;
    * Verbal position reporting.

7) Expand upon your answer to the question above in some detail using the following terms:

   * Radar echo;
   * Interrogator;
   * Transponder;
   * Transmission;
   * Mode;
   * Code.

8) ATC based on verbal position reports, outside the range of surveillance radar is known as procedural control. Although safe, why is procedural ATC less efficient than ATC based on radar surveillance.

9) Why is ADS sometimes described as a pseudo radar?

10) Initially, ADS will most likely be used in conjunction with procedural ATC; that is, controllers will watch the ADS display but continue to apply procedural separation standards. It is expected however, that ADS will eventually allow for a reduction of the procedural separation minima in areas where ADS is employed. What issues do you think have to be resolved before a reduction in the separation minima can be implemented using ADS.

11) Explain how information from the FMS on board aircraft, automatically transmitted to the ground using ADS, could be used by an automated ATM system?

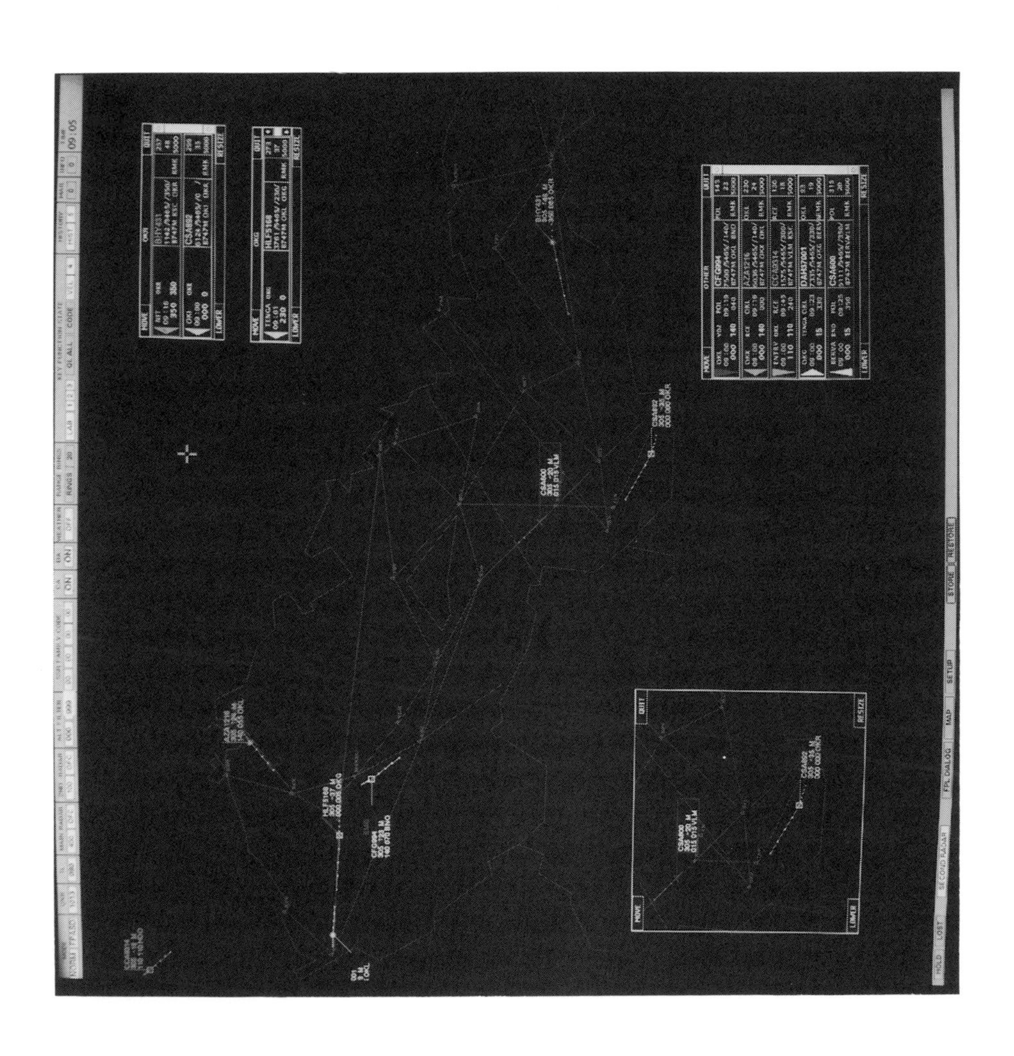

An enhanced ATM display designed to ease controller workload. Electronic flight strips replace conventional paper strips for controllers to store data. (Picture provided courtesy of Thompson CSF.)

6. Air Traffic Management

Introduction

> A general objective of ATM is to enable aircraft operators to meet their planned times of departure and arrival and adhere to their preferred flight profiles with minimum constraints without compromising agreed levels of safety (ICAO. 1991a, p. 5-1).

Future ATM systems will strive toward the objective stated above. That is, to give aircraft operators the freedom to dynamically allocate their preferred flight profiles. Many States and regions are already developing their systems to accomplish this objective. The *Free Flight* concept of the United States is one example of the drive toward more autonomy of flight. The European Air Traffic Management System (EATMS) is another. These concepts are explained in more detail later in the chapter.

The primary objective of air traffic control is, and will continue to be, centred on the need to ensure the maximum degree of safety. Air traffic control functions to prevent collisions while expediting and maintaining a safe and orderly flow of traffic. The order of these attributes can never be rearranged (Ruitenberg, 1995).

As envisaged by the world community at the ICAO Tenth Air Navigation Conference, and further elaborated upon in the work programme of the ICAO Air Navigation Commission, ATC should continue to evolve into an integrated, global ATM system. In addition to ATC's primary objectives of safety and efficiency, some other objectives to be reached through the envisaged ATM system can be summarized as follows (ICAO, 1993a):

* To meet evolving air traffic demand;
* To support a safe and orderly growth of international civil aviation;
* To enhance safety, regularity and efficiency;
* To enhance economy of commercial air transport;
* To optimize benefits through global integration.

In order to reach these objectives, it is necessary to identify target system enhancements and benefits, to solidify planning on a global, regional and national basis, and to manage the ATM system evolution with a view to obtaining progressive and incremental enhancements, functional compatibility of all of the elements of the system, including flight operations and, ultimately, integration into a global system (Heijl, 1994).

In this context, CNS systems and associated emerging technologies should not be seen as ends in themselves, but rather as tools that will enable ATM enhancements which will result in benefits for both airspace users and providers of services. ATM should therefore dictate the operational requirements for CNS systems and their functionality. Based on this premise, optimum benefits from CNS systems will only be obtained through an integrated, global ATM system.

In reexamining the benefits expected of the CNS components, briefly, we recall that implementation of these systems will improve the handling and transfer of information by making extensive use of data link; improve navigational accuracy through global navigation satellite systems; and, extend and improve surveillance capabilities using automatic dependent surveillance and Mode S radar.

It is the evolutionary implementation of the CNS elements and their orchestrated interaction, along with the reasonable use of automation to assist controllers in performing a part of their cognitive tasks, that will form the backbone of an integrated global ATM system.

A significant portion of today's air carrier aircraft operate with quite sophisticated equipment with on-board capabilities that generally far exceed what is presently used by ATC systems individually. For this reason, the International Air Transport Association (IATA), on behalf of the airlines, has strongly endorsed the implementation of CNS/ATM systems technology

(1991). The airlines recognize the benefits that CNS can bring to ATM, which would allow them substantial savings through reduced operating costs and increased opportunities to operate in conformance with the most efficient flight profiles. As stated above, this is the general objective of ATM. Furthermore, a more efficient use of the airspace would lead to a subsequent reduction in delays.

The FANS Committee and the Tenth Air Navigation Conference provided a broad background on how the CNS elements should be brought together to form an integrated, global ATM system. The ICAO Air Navigation Commission has since expanded upon that framework and has developed a structured work programme for ICAO's Air Navigation Bureau and the regional planning groups to follow. Elements are beginning to be put in place to varying degrees in the different ICAO regions and by individual States. The goal of ICAO now is to develop a coherent strategy that will bring all of the elements together to form an integrated, global system.

This chapter reviews the ICAO ATM work programme along with ATM developments taking place around the world. Much of this work is still in the planning stages.

This review also includes an examination of the various elements of the ATM system and some of the research, trials and demonstrations being conducted.

Definitions

In order to fully understand this chapter, it is necessary to first become familiar with a few of the common international terms which are frequently used when referring to ATM and which you will see often as you become involved with international civil aviation.

Airborne Collision Avoidance System (ACAS) An aircraft system based on secondary surveillance radar (SSR) transponder signals which operates independently of ground-based equipment to provide advice to the pilot on potential conflicting aircraft that are equipped with SSR transponders.

Air traffic control service A service provided for the purpose of:

a) preventing collisions:

1) between aircraft, and

2) on the manoeuvring area between aircraft and obstructions; and,

b) expediting and maintaining an orderly flow of air traffic.

Air traffic service (ATS) A generic term meaning variously, flight information service, alerting service, air traffic advisory service or air traffic control service (aerodrome, approach, enroute).

Aerodrome control service Air traffic control service for aerodrome traffic.

Approach control service Air traffic control service for arriving or departing controlled flights (within a given radius of the aerodrome).

Area control service Air traffic control service for controlled flights in controlled airspace (enroute).

Flight information service A service provided for the purpose of giving advice and information useful for the safe and efficient conduct of flight.

Flight plan Specified information provided to air traffic services units, relative to an intended flight or portion of a flight of an aircraft.

Shortcomings of the present system

Previous chapters described the shortcomings of present systems as identified by the FANS Committee, which recognized that for an ideal worldwide air navigation system, the ultimate objective would be to provide a cost-effective and efficient system that would permit the flexible employment of all types of operations in as near four dimensional freedom as their capability would permit (ICAO, 1988b).

The list of shortcomings of present ATC systems below is applicable, to some extent, to even the most developed ATC system environments:

* The information flow within and between ATC units (ground-ground communications) and between ATC units and aircraft under their control (air-ground communication) is insufficient to support further significant improvements;

* ATC needs improved data and procedures for surveillance, prediction and optimization of the air traffic flow;

* The most advanced ATC systems continue to work on the basis of data representing the aircraft performance and environment conditions which only poorly approximates reality. Therefore, only limited accommodation of optimized flight profiles is achieved;

* The capability of advanced airborne equipment in the field of planning and optimization of flight paths has outstripped that of the ground systems to support it. Operators are pressing to be able to more fully exploit such capabilities. A way to accommodate aircraft will be through implementation of concepts which take these capabilities more fully into account; and,

* Route structures are generally inflexible.

Furthermore, the conventional airspace organization of flight information regions and their supporting infrastructure of ATS routes and ground-based facilities and services is largely based on national requirements and sovereignty, rather than on international requirements. As a result, fragmentation of the airspace, and a diversity of national systems prevent an optimum use of the airspace (ICAO, 1995i).

Finally, because of the inherent limitations of air traffic control systems to separate numerous aircraft on random routings, and a lack of automation to assist with conflict detection and resolution, aircraft must plan their flights along routes and be channelled, to a certain degree, in order for the air traffic control system and the people that operate it, to safely keep aircraft separated from each other.

Air Traffic Management (ATM)

> In combination, the new CNS system provided under the ICAO concept will make it possible to realize a broad range of ATM benefits that will enhance safety, reduce delays, increase capacity, enhance system flexibility and reduce operating costs (ICAO, 1993a, p. 8B-4).

The chapters in this book dealing with communication, navigation and surveillance comprised a review of the various types of equipment and the associated technological developments that will form the tangible part of the future system. These developments will all serve to make a more efficient and effective ATM system. Unlike CNS technical systems, hardware and software, however, when referring to ATM, there is no readily identifiable piece of equipment that can be seen as revolutionary, although certain flight data and radar data processing systems and associated software, specific to ATM, will form and integral part of future CNS/ATM systems. ATM is rather a system of rules and procedures, and air traffic controllers and future air traffic managers apply those rules and procedures to achieve a safe and efficient system (ICAO, 1991a).

The new technologies and aircraft capabilities will require evolutionary changes to these rules and procedures and the overall approach to them, because the new capabilities will permit far more cooperative arrangements than before. Furthermore, the CNS technologies described throughout this text can only be fully exploited if there is international harmonization of ATM standards and procedures. If international operators improve their on-board capabilities to exploit ATM service improvements, they will expect a return on their investment that can only be had if the same improvements are implemented in many States. From the aircraft operator's point of view, it must be possible to equip aircraft operating internationally with a single set of avionics usable everywhere. Additionally, many of the expected service improvements cannot be meaningfully implemented by one State, but must be implemented in contiguous regions through which a significant number of aircraft will fly (ICAO, 1991a). Figure 6.1 gives an example of a complete set of CNS/ATM equipment.

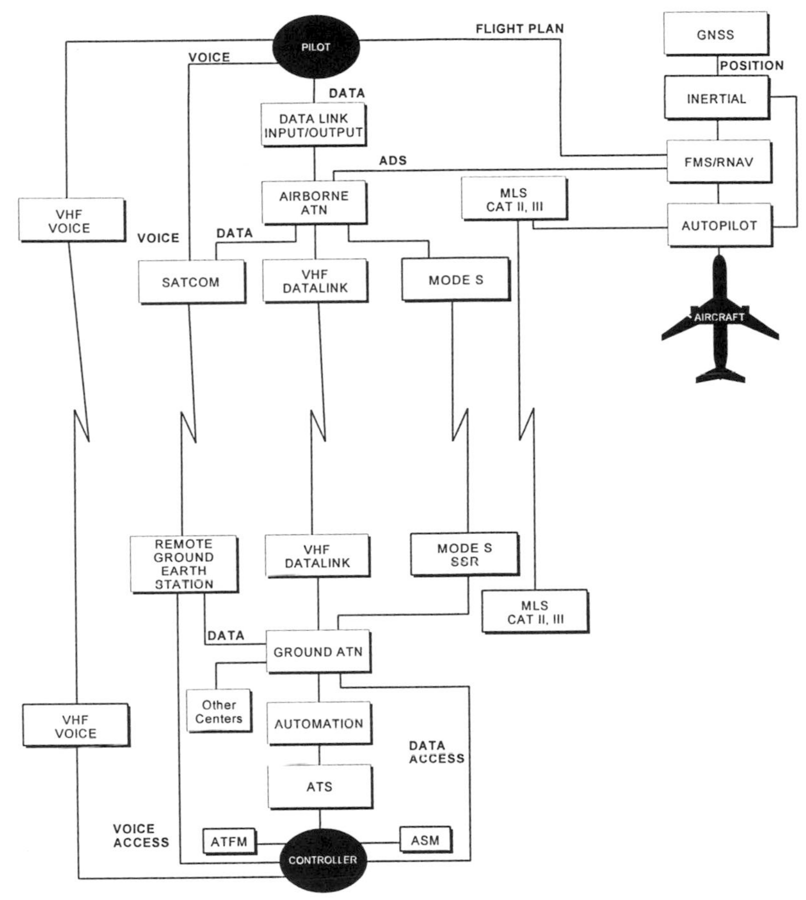

Figure 6.1 Example of a complete set of CNS/ATM equipment

Goals

The goals for the future ATM system include the following (ICAO, 1991a):

* Maintenance of, or increase in, the existing levels of safety;

* Increased system capacity and full utilization of capacity resources as required to meet traffic demand;

* Dynamic accommodation of user-preferred three-dimensional and four-dimensional flight trajectories;

* Accommodation of a full range of aircraft types and airborne capabilities;

* Improve provision of information to the users such as weather conditions, traffic situation and availability of facilities;

* Improve navigation and landing capabilities to support advanced approach and departure procedures;

* Increase user involvement in ATM decision-making including air-ground computer dialogue for flight negotiation;

* Create, to the maximum extent possible, a single continuum of airspace, where boundaries are transparent to users; and,

* Organize airspace in accordance with ATM procedures.

ATM has been referred to often in this book as an element in CNS/ATM systems. It is important now to understand the procedural components that constitute ATM and how the orderly development of and agreement on worldwide standards, procedures and operational requirements will allow these components to be planned and implemented to form an integrated, global ATM system.

Operational concept

Attaining the goal of an integrated, global air traffic management system, requires harmonization and standardization of regional and national system elements of equipment and procedures. ICAO, through its Air Navigation Commission, the Secretariat, panels and other bodies, is developing new standards and recommended practices that will form a part of the Annexes to the Convention. States and industry then use these standards as a guide in development and implementation of ATM systems.

A structured work programme for ICAO, dealing specifically with the integration of the future ATM system has been defined. That programme has been agreed to by the Air Navigation Commission and approved by the ICAO Council (see Figure 6.2).

That ATM work programme recognizes that before developing the standards necessary for harmonization and integration, an operational concept of the future ATM system first needs to be defined. This will clarify the benefits and give States and industry a clear objective for designing and implementing ATM systems. The issues being addressed by ICAO, and which will become a part of the operational concept leading to standards and procedures, are:

* Autonomy of flight;
* Separation assurance;
* Situational awareness;
* Collision avoidance;
* Optimization of traffic flows;
* The regional concept of providing ATM services, covering greater geographical areas, encompassing several Flight Information Regions (FIR); and,
* Required Total System Performance (RTSP).

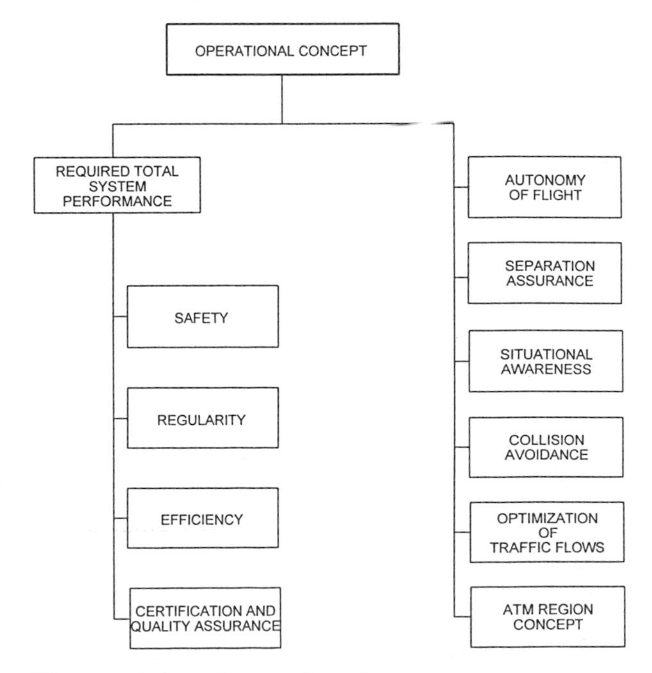

Figure 6.2 Operational concept

Required Total System Performance (RTSP) Although the issues mentioned above are not new, except for the concept of RTSP, they will have to be addressed in the context of an integrated system. RTSP on the other hand, will need to be clearly defined as it is a cornerstone for sound planning and development of future systems.

ICAO has developed worldwide standards for many aspects of civil aviation, however, the current ATM system has evolved without globally agreed criteria for safety, regularity and efficiency of international civil aviation having been established. A target level of safety has been defined only for some airspaces, but not on a global level. In the absence of agreed criteria for airspace/airport capacity and for flexible use of airspace, there is no common basis for regularity and efficiency worldwide. As a result, there is no assurance that the future traffic demand and airspace users' needs can be met (ICAO, 1995j).

In the context of the above, the future system must be viewed in its totality. The total system can be seen as the totality of airspace, flight operations, and the facilities and services provided. When fully defined, RTSP will include criteria that should be met by the entire ATM system in the areas of safety, regularity, efficiency, sharing of airspace and in the area of human factors.

Theoretically, criteria could be developed in terms of acceptable delays, and efficiency of flight operations could be expressed in terms of mileage ratio between actual routes flown and great circle routes for example. Certification of systems and procedures will also need to be considered, as well as any necessary measures to provide for ongoing quality assurance. In this respect, the ICAO Flight Safety and Human Factors programme is investigating the possibility of certification of systems based on human factors requirements.

RTSP would allow the provider and users of a given airspace to determine the optimum way that the airspace could be used. For example, lower performance standards could be acceptable in a particular airspace, for some or all system components, if the users were prepared to accept larger separation standards.

The operational concept and associated RTSP, when fully defined, would serve as a benchmark for regional air navigation planning, and will be included as a formal part of the regional air navigation plans (see Chapter 2 for a description of regional air navigation plans), offering guidance to the ICAO regional planning groups that carry out the actual planning of infrastructure that serves international civil aviation.

Elements of the ATM system

The envisaged ATM system will consist of several subelements; these are: flight operations, ASM, ATS and ATFM. In the future ATM system these subelements will evolve and take on different roles than exist today, mainly because they will integrate into a total system. And rather than view air and ground as separate functions, flight operations will be fully integrated as a functional part of the ATM system. Ultimately, this interoperability and functional integration into a total system will yield a synergy of operations that does not currently exist. In fact, the capabilities of airborne and ground-based systems cannot be fully exploited in the absence of functional integration. Data link will be used for data exchange between elements of the ATM system thereby acting as the medium whereby functional integration will be accomplished.

The paragraphs below describe the elements of the future ATM system as envisaged, along with the expected changes that will permit the functional integration and interoperability necessary for successful integration of that system.

Flight operations

The airborne and ground parts of CNS/ATM systems will interact with each other in new ways. For example, automated systems on the ground will assist the controller with conflict detection and resolution, based on information downloaded from aircraft flight management systems and at some point, will negotiate ATC clearances with these airborne systems. Additionally, other information that is now accomplished by voice, will increasingly be carried out using automatic transmission of data.

Recognizing the significance of the airborne component to ATM, the requirements for the ATM functional capabilities of systems such as airborne collision avoidance systems (ACAS), flight management systems (FMS) and airborne data bases are being developed. In this context, the ADS Panel is currently developing requirements related to aircraft interface of ADS and ATS data links (see Chapter 5). The Human Factors Study Group has also begun work on identifying and addressing the specific human factors related problems associated with CNS/ATM systems in general and the integration of the elements and the effects on the human operators in the system. This aspect is addressed in more detail in Chapter 8.

Additional work and identification of tasks dealing specifically with flight operations will be developed as the need is identified. For example, as work on ADS-Broadcast (see Chapter 5) matures, the interaction between the ground and airborne systems, as well as between airborne systems, will have to be addressed in the context of an integrated and fully functional ATM system.

Airspace management (ASM)

Airspace management is traditionally recognized mainly as involving a dynamic or tactical sharing of airspace by civil and military users. In the future ATM system, however, ASM will not be limited only to tactical aspects of airspace use. Given a revised definition and function, ASM will form a cornerstone in the future ATM system. Its main scope will be toward a strategic planning function of airspace infrastructure, and flexibility of airspace use.

In relation to ATM, ASM is seen to consist of two main elements: first, the determination, for any given airspace, of the ATM requirements for communications, navigation and surveillance; and second, infrastructure planning.

These elements are described below.

ATM requirements for communication While ATS communication requirements are established in ICAO Annex 11 and technical specifications for various communications systems are established in Annex 10, there are not yet formal requirements to specifically support the future ATM system. ICAO is therefore developing the ATM requirements for communications that will specify the air-ground and ground-ground communication coverage required for both voice and data link needed for the development of an integrated, global ATM system which would be supported by the technological developments outlined in Chapter 3 (e.g., satellites, data link, ATN) (ICAO, 1995k). This work will include a statement of required communication performance (RCP), a new term that will parallel and complement the work already accomplished concerning required navigation performance (RNP) and work still to be done regarding required surveillance performance (RSP).

These requirements when completed, will govern the development of future Standards and Recommended Practices (SARPs), procedures and guidance material for communication system performance and equipment carriage requirements for aircraft. This will guide the States and industry in the design, development and implementation of communications systems both in the air and on the ground, based on CNS/ATM systems.

ATM requirements for navigation The ATM navigation requirements will specify a need for area navigation (RNAV) capability by all aircraft, which would be supported by the technological developments outlined in Chapter 4 (e.g., GNSS). While initially this RNAV capability may be provided by airborne systems which rely on ground-based navigation aids, there will be a shift towards GNSS based systems. This will lead to one of the main economic benefits of CNS/ATM, which is the eventual withdrawal of the current ground-based short-range navigation aids.

ICAO is developing the ATM requirements for navigation capability and performance for enroute and terminal area operations, which will govern the development of necessary SARPS, procedures and guidance material on navigation capability (ICAO, 1995l). Work will continue on developing the concept of required navigation performance (RNP). This material will guide States and industry in the design, development and implementation of navigation systems in the air and their supporting infrastructure on the ground, based on CNS/ATM systems.

ATM requirements for surveillance Without additional radar and/or ADS, airspace capacity will be insufficient to accommodate future air traffic demand with necessary efficiency. The ATM surveillance requirements will specify criteria as to where radar and/or ADS coverage is required for ATM purposes, which would be supported by the technological advancements outlined in Chapter 5 (e.g., Mode S, ADS).

ICAO is developing the ATM requirements for surveillance coverage (ICAO, 1995m). This work will also include a statement of required surveillance performance (RSP) to parallel the work already accomplished concerning RNP and to complement the work yet to be done regarding RCP. These requirements will govern the development of SARPS, procedures and guidance material leading toward the global sharing of surveillance data from ADS, SSR and integrated ADS/SSR systems. This material will guide States and industry in the design, development and implementation of surveillance systems with components both in the air and on the ground, based on CNS/ATM systems.

Infrastructure planning Infrastructure planning is the second main element of ASM. Conventional airspace planning of flight information regions and their supporting infrastructure of ATS routes and ground based facilities and services has largely been based on national requirements and sovereignty, rather than on international requirements. As a result, fragmentation of airspace, and a diversity of national, conventional systems prevents optimum use of the airspace.

The efficiency objectives of future global ATM will not be realized unless the airspace infrastructure is developed in a harmonized manner and in accordance with global efficiency requirements and safety levels. Future ATM therefore requires a cooperative approach. Airspace infrastructure planning, if applied over larger geographical areas than is currently the practice, will offer greater benefits.

ICAO is developing the operational requirements and planning criteria for airspace organization, services and facilities to support ATM, on the basis of an airspace planning methodology, with the objective of facilitating the optimal use of airspace, organized so as to provide for efficiency of service, while maintaining or improving the existing levels of safety (ICAO, 1995n).

The RGCSP has developed an airspace planning methodology to be used as a tool, using risk assessment modelling, to derive safe separation minima for use in a given airspace, on the basis of volume of traffic, facilities and services, as well as on airborne capabilities. This tool provides implementation options for ground-based facilities and airborne systems to achieve required functionality, on the basis of stated objectives in terms of aircraft movements and operationally desirable separation minima.

From the above paragraphs, the evolving and significant role of the function of ASM can be seen. To take advantage of the benefits of CNS in the future ATM environment, it is necessary to determine the requirements for communications, navigation and surveillance so that these elements can be developed, planned and implemented in the most cost-efficient, effective, yet integrated way possible.

Air Traffic Services (ATS)

ATS will continue to be the main element of ATM. ATS itself is comprised of several subelements, one of these being air traffic control (ATC). The safety and efficiency of air traffic is achieved primarily through ATC and the other elements of ATS [i.e., flight information service, alerting service, air traffic advisory service and air traffic control service (aerodrome, approach, enroute)].

Provisions are currently being developed by the ADS Panel which will govern the development of standards, procedures and guidance material necessary to guide States and industry in the design, development and implementation of ATC systems based on ADS with components both in the air and on the ground. Work on integration will ensure that these systems are developed in a manner so as to provide harmonization and integration into a regional and global network of continuous service.

There are several direct benefits that will come to ATS with the implementation of CNS/ATM systems as follows:

- A safety benefit through reduction in ATC communication errors will be achieved through the use of data link and clear voice channels over areas where there is presently no direct pilot controller communication availability;
- More optimum routings without the need for ATC vectoring can be obtained through better navigation based on the use of GNSS;
- Blunder detection and tactical control flexibility will be afforded through global surveillance based on ADS;
- Enhanced safety and increased capacity, as well as a reduction in human errors can be expected to result from automation support, based on sound human factors principals, for both controllers and pilots;
- Efficiency and regularity of international civil aviation can be improved by automated interfaces between adjacent ATC units;
- Optimum flight profiles and better economy of flight can be achieved by interfacing airborne and ground systems.

Air Traffic Flow Management (ATFM)

The purpose of ATFM is to optimize air traffic flows, reduce delays to aircraft both in flight and on the ground, and prevent system overload.

ATFM can be subdivided into two categories: strategic and tactical. Strategic ATFM seeks to accommodate traffic through preplanning such as by the use of traffic routing schemes for particular areas or during particular seasons. Additionally, predetermined acceptance rates for specific points over the route of flight, or pertaining to certain destination airports may be prescribed on a strategic basis. Acceptance rates are usually enforced through the issuance of departure slot allocation times.

Tactical ATFM on the other hand, attempts to deal with constraints on a flexible, real-time basis.

The increasing number of long-range interregional flights may require integration of regional ATFM systems and procedures. To ensure global compatibility of regional ATFM systems, standardization of functionality is required. Such standardization is being undertaken as part of the technical work programme of ICAO through the development of functional specifications and procedures for the worldwide integration of ATFM systems, which would facilitate an optimal flow of air traffic (ICAO, 1995p; ICAO, 1995q; & ICAO, 1995r).

As part of the work programme, the following areas are being addressed:

* ATFM related data bases;
* User interface with ATFM systems;
* Surveillance input for ATFM systems.

Procedural areas that form part of the work programme regarding ATFM are:

* Strategic ATFM procedures;
* Tactical ATFM procedures;
* Slot allocation procedures.

In order to assist a possible ATFM system integration into a global ATFM network, the following areas are being addressed:

* Global networking;
* Integration of regional ATM systems and procedures.

Interoperability and functional integration of flight operations, ATS, ATFM and tactical ASM

The systems described above have traditionally been viewed as independent systems forming part of the whole ATC system. As specified in so many places in this text, the functional integration of all of the elements of the ATM system is the key to its success. For this reason, ICAO has made the aspect of integration, the subject of a particular task. That task is precisely to determine the ATM requirements and functional specifications for integration of ground-based and airborne systems. ICAO is therefore developing the SARPS, procedures and guidance material necessary for the functional integration of airborne and ground-based systems (ICAO, 1995o). This material will guide States and industry in the design, development and implementation of all of the elements of the ATM system, based on CNS/ATM systems, leading toward an integrated system.

Increasing numbers of aircraft are equipped with new technology CNS systems which provide significant capability to proceed along any desired flight path. The FANS packages being installed by Boeing, Airbus and McDonnell Douglas described in previous chapters are examples of these systems. Current supporting ATS systems of varying capability do not permit optimum flight trajectories in most airspaces. The capabilities of airborne and ground-based systems cannot be fully exploited in the absence of interoperability and functional integration of these systems (see Figure 6.3).

ATM Automation

> The need to put many more aeroplanes safely in any given airspace sector....can be achieved only by pre-programming air traffic control computers with the aircraft's flight plan, and having the aircraft's flight management system continually feed back its performance, position and the crew's intentions through datalinks (Air Transport, 1995, p. 27).

Air traffic control systems worldwide are undergoing some of the boldest modernization efforts in history. The computer-human interface is changing significantly with the introduction of new hardware and software. Electronic displays are replacing the paper flight strip, colour is being introduced to radar displays, along with shading, blinking and inverse video, as well as new technologies such as icons, menus and pointers (Hopkin, 1989). New

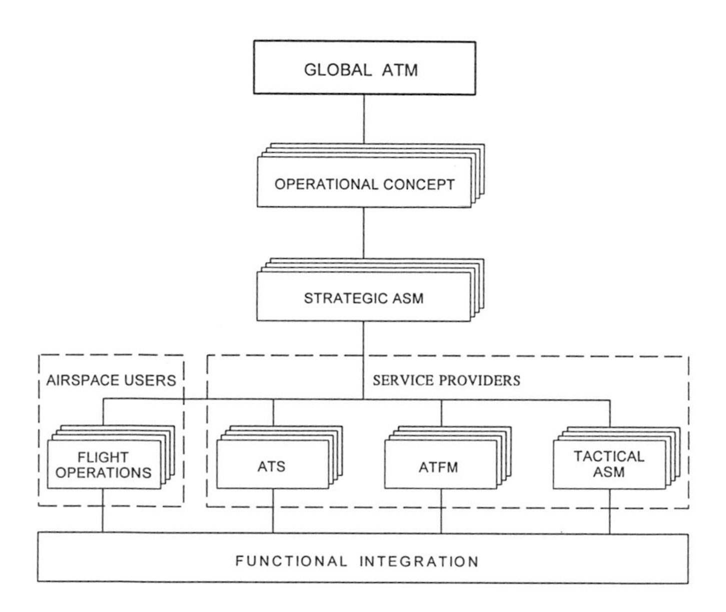

Figure 6.3 A structured approach to the work on global ATM

types of entry devices are being added to new communications systems which are changing the way that controllers communicate with aircraft by taking advantage of data link. Most significantly, however, will be the introduction of software that will perform a part of the cognitive tasks that the controller now performs.

The air traffic controller's job consists of complex tasks demanding a high degree of skill and active application of unique cognitive abilities such as spatial perception, information processing, reasoning and decision-making (Della Rocco, 1990). The controller must know where all of the aircraft under his/her responsibility are, and determine how and when to take action to ensure that they remain separated from each other, while also seeing to their requests and needs for descent, climb, takeoff, departure, etc.

Although it is well accepted that the human controller in the system has performed these tasks more than adequately over the years, it is also accepted that there is room for improvement, much of which will come from automation in the form of advanced software.

Some of the names commonly associated with this kind of software are intelligent processing, artificial intelligence and automated reasoning (Galotti, 1991). The justification for implementation of these systems is that the controller is not always capable of determining exactly when and/or where aircraft would conflict with each other, so he/she must take action whenever there is doubt. Quite often, this can lead more to a separation of blocks of airspace rather than of individual aircraft. Added together over an entire system, this results in less than an optimum use of the airspace (Galotti, 1996).

The expectation is that greater degrees of accuracy could be achieved through the rapid calculations associated with automation. As more exactness is built into the system, there would be a corresponding decrease in the need to issue ATC clearances, leading to an increase in airspace capacity. Automated reasoning could at first predict ATC problems based on known aircraft intentions, but could eventually add complex resolution processes. These systems will be introduced in an evolutionary manner as technology and reason will permit. They are expected to increase airspace capacity by assisting the controller in carrying out, and in some cases assuming, the advanced cognitive tasks that he/she now performs.

There are several issues that need to be carefully addressed when considering automation of this nature. The most critical is based on the fact that aircraft do not always do what they are expected to do, whether because of poor communications, emergencies, or blunders. The human controller, for all his/her inefficiencies, is also very flexible and adaptive and quite capable of compensating and/or developing alternative plans. Based on this, it is reasonable that computers and software will assist the controllers in accomplishing at least initially, only a part of their cognitive tasks. It is also unrealistic to determine at this early stage, that computers could effectively replace controllers in the near term, mainly because of their uniqueness in providing the aviation system a degree of flexibility that cannot now and may never be attained by computational systems. For these reasons, the ICAO Human Factors Digest on CNS/ATM systems (1994i) states that:

> automation should be viewed as a tool or resource, a device, system or method which enables the human to accomplish some task that might otherwise be difficult or impossible, or which the human can direct to carry out more or less independently a task that would otherwise require increased human attention or effort (p. 3).

In this light, automation is seen as one of many resources available to the human operator, who retains the responsibility for management and direction of the overall system. Additionally, unexpected or unplanned events must be a required part of planning and design when considering the automation systems that would replace the cognitive activities of controllers.

Based on this approach, technology and automation should not be seen as an end in themselves, however, applied in a judicious and evolutionary way, can alleviate some capacity constraints and improve the system overall.

The introduction of new CNS capabilities and increasingly capable ground and airborne systems, will allow the evolution of sophisticated ATM systems. The future ATM system will make increasing use of automation to reduce or eliminate a variety of constraints imposed on ATM operations by current systems, and to derive the benefits made possible by implementation of the new CNS systems in the following ways:

* ATM automation will make it possible to formulate real-time flow management strategies;
* ATM automation will allow negotiation between ATM and aircraft to enhance tactical control;
* Data link and voice channels, enhanced by automation aids, will be used for aircraft not capable of automated negotiation with ATM;
* Flexible oceanic ATM will accommodate user-preferred trajectories;
* Both flow management and tactical control will be enhanced for enroute and terminal operations;
* Traffic is expected to flow smoothly into and out of terminal areas;
* Air-ground information exchange will be greatly improved; and,
* New automated ATM ground systems will support increased capacity (ICAO, 1993a).

In areas of high density traffic it will be necessary to implement powerful automation systems to respond to the ever increasing demand. The use of ATM automation tools will help air traffic controllers and the ATM system as a whole to sort, process and display information. This will assist air traffic managers to visualize the data and consider the range of alternatives (Andresen, 1991).

Summary of the benefits of the future ATM system

* The new ATM capabilities and more accurate data will make it possible to enhance safety, reduce delays, and increase airspace and airport capacity;

* Oceanic ATM operations will become much more flexible, resulting in a greater capability to accommodate user-preferred trajectories;

* Improved flow management will prevent excessive levels of congestion;

* Improved tactical control will maximize dynamic accommodation of user-preferred three and four dimensional trajectories in meeting ATM constraints;

* Data link will transmit a variety of information from appropriately equipped aircraft to the ground, and provide enhanced information to the cockpit. This will dramatically decrease the communicator's workload;

* New capabilities will make it possible to permit flexible routing, as well as dynamic modifications to aircraft routes in response to changes in weather and traffic conditions;

* Terminal and enroute ATM functions will be integrated to provide smooth traffic flows into and out of terminal areas;

* Air traffic controllers or managers will be able to establish efficient approach streams for parallel and converging runway configurations;

* Single-runway capacities in instrument meteorological conditions (IMC) will increase to a level approaching current single-runway capacities in visual meteorological conditions (VMC);

* Independent instrument flight rules (IFR) operations on triple and quadruple parallel runways will become routine in high density airspace;

* Conflicts among departure and approach operations involving adjacent airports will be reduced;

* Flexibility in controlling the noise footprint of airport traffic operations will be increased (ICAO, 1993a). Figures 6.4 and 6.5 give an overview of the benefits expected to ATM from the evolutionary implementation of CNS systems.

A glance at the future

The civil aviation community is at a historic standpoint. Because of the inabilities of current ATC systems to support the growing air traffic demand, a new concept has been developed based on advanced technologies. But technology alone cannot solve the problem. It is the extent to which the technology is used to support an integrated, global ATM system, that will determine the success of CNS/ATM systems. Furthermore, for maximum advantage to be had from the CNS/ATM systems technologies, the world community must be willing and able to implement and adhere to internationally agreed upon standards and procedures, designed to allow the evolutionary implementation of the new ATM system, and should then proceed with implementation of systems and procedures across large, homogeneous, geographic areas. It is for this reason, that this text has explained the historic significance of ICAO, its planning mechanisms and of international standardization and harmonization. It is precisely through this international machinery that CNS/ATM systems came into being and through it that ATM may be successfully implemented.

A cooperative approach to achieve the objectives of safety, regularity and efficiency of services is needed. This basic planning principle has never been more true than at the present time as we plan for the future ATM system.

	ATM ELEMENTS CONSIDERED	ATM FUNCTIONS						
		Strategic conflict detection	Tactical conflict detection	Conflict resolution	Accommodation of pilot request	Civil/military co-ordination	Airspace sharing (ASM)	Capacity management (ATFM)
PRESENT	**PROCEDURAL** (Non-radar) Basic environment	Flight progress strips	Flight progress strips	Manual (altitude changes)	Pilot/controller co-ordination	Voice	Airspace segregation	En route + Departure slots
PRESENT	**RADAR** Environment	Flight progress strips	Radar + Short term conflict alert (strips as back-up)	3 D* Radar vectors + Altitude change + Speed restriction	Pilot/controller co-ordination 3D*	Computer linkages + Voice	Temporary segregation + Tactical co-ordination by telephone possibilities	Departure slots
FUTURE	**FANS** Environment	Computer aided conflict prediction and resolution	Radar and/or ADS + Short term conflict alert + Suggestions for solutions	4 D* Radar vectors + alt. changes + speed restriction + time over fixes	Exchange FMS/ATC data pilot/controller agreement 4D*	Computer linkages + Voice and area concept	Area concept + Flexible use of airspace	Flight plan data base + Computer assisted slot allocation + Re-routing procedures
FUTURE	**BENEFITS** To ATM	• More efficient use of airspace • Resolution with minimum disruption of flight profiles	• More efficient use of airspace and A/C capabilities	• More flexibility • Better flight profiles • Enhanced safety	• More efficient flight profiles • More dynamic flight planning	• Airspace capacity increase	• Airspace capacity increase	• Faster response • Less delays • Better use of available capacity

* 1D e.g. vertical
3D e.g. vertical, lateral, longitudinal (speed restrictions)
4D e.g. vertical, lateral, longitudinal (speed restrictions), time (at a fix)

Figure 6.4 Benefits expected to ATM from the evolutionary implementation of CNS systems in continental airspace

	ATM ELEMENTS CONSIDERED	ATM FUNCTIONS				
		Strategical conflict detection	**Tactical conflict detection**	**Conflict resolution**	**Accommodation of pilot request**	**Capacity management (ATFM)**
PRESENT	**PROCEDURAL** (Non-radar) Basic environment	Manual Automatic (where available)	Manual Automatic (where available)	Manual	Pilot/controller co-ordination	Organized track system Interface with adjacent FIRs
FUTURE	**FANS** Environment	Computer Aided Conflict Detection and Resolution	• Short term conflict alert (ADS) • PLN non-conformance alert	4D* (altitude changes + Speed restr. + heading changes time over fixes)	Exchange FMS/ATC data Pilot controller agreement 4D*	Computer aided conflict detection ADS
FUTURE	**BENEFITS** To ATM	• More efficient use of airspace • Resolution with minimum disruption of flight profiles • Less controller workload	• More efficient use of airspace and A/C capabilities	• More flexibility • Better flight profiles • Enhanced safety • Less controller workload	• More efficient flight profiles • More dynamic flight planning • Less controller workload	• More efficient flight profiles • Increased airspace capacity

* 1D e.g. vertical
3D e.g. vertical, lateral, longitudinal (speed restrictions)
4D e.g. vertical, lateral, longitudinal (speed restrictions), time (at a fix)

Figure 6.5 Benefits expected to ATM from the evolutionary implementation of CNS systems in oceanic airspace

The chapters in this book dealing with the individual elements of communication, navigation and surveillance describe much of the research, trial, development and implementation that is being accomplished around the world by States, industry, ICAO and its regional planning groups.

Taken together, these offer a useful compilation and view toward the future of CNS/ATM. A truly integrated ATM system will only be possible after the standards and recommended practices, procedures and guidance material are developed, adhered to and implemented by States. This system does not yet exist. It will probably never exist in its ideal state however, substantial progress should be made in that direction. Advances are being made by individual States, ICAO regional planning groups and by industry.

You should now review the sections "A glance at the future", in all of the previous chapters and determine how developments can lead to the ATM system described in this chapter.

CNS/ATM related work of regional planning and implementation groups (ICAO, 1995c)

African Planning and Implementation Regional Planning Group (APIRG) An APIRG CNS/ATM task force has developed a CNS/ATM implementation plan for that region which includes a framework, strategy and implementation tables, based on major routes (ICAO, 1995b). The ninth meeting of APIRG was held in February 1995. To promote CNS/ATM in Southern African States, a three day CNS/ATM seminar was conducted in Pretoria, South Africa in April 1995. This seminar preceded the meeting of Directors General of Civil Aviation and Chief Executive Officers of airlines of the Southern African Development Community (SADC) which was then apprised of the potential of CNS/ATM to resolve the shortcomings of the present air navigation systems and of the economies and benefits that could be realized through the sharing of resources by those States. The ICAO concept was widely accepted by SADC and actions were initiated towards its planning and implementation (ICAO 1996).

Asia/Pacific Air Navigation Planning and Implementation Regional Group (APANPIRG) APANPIRG developed an Asia/Pacific Regional Implementation Plan for new CNS/ATM systems and proposed the development of an action oriented implementation workshop programme to support early implementation of CNS/ATM systems, as well as the development of guidance material to promote early application of GNSS technology. Nine geographical areas have since been established to accommodate major traffic flows in the Asia Pacific Regions, where ATM benefits could be easily identified for early implementation of CNS/ATM systems (ICAO, 1996).

Caribbean/South American Regional Planning and Implementation Group (GREPECAS) GREPECAS has recommended that States in this region with similar air traffic and airspace interactions be encouraged to conduct joint CNS/ATM tests and trials with a view to substantiate the benefits that can be derived from the early introduction of CNS/ATM systems; that States develop their own plans for transition to CNS/ATM systems and that ICAO study the possibility of holding seminars on the institutional and legal aspects associated with CNS/ATM systems. That group also approved a regional CNS/ATM transition plan. The development of a Caribbean, South American Regional Implementation Plan for CNS/ATM systems has been significantly advanced and approved by GREPECAS. In order to be more responsive to advancing phases of the project, it has been decided to establish a CNS/ATM Implementation Coordination Subgroup of GREPECAS (ICAO 1996).

European Air Navigation Planning Group (EANPG) The EANPG developed a CNS/ATM plan and considered the processes and global institutional arrangements for its implementation. This group has also extensively covered the subject of the impact of GNSS operations on aerodrome operations and instrument flight procedures. The Thirty Seventh Meeting of the EANPG created a subgroup to coordinate the work of the various EANPG bodies concerning CNS/ATM (ICAO, 1996).

Middle East Air Navigation Planning and Implementation Regional Group (MIDANPIRG) MIDANPIRG established a CNS/ATM sub-group to review, monitor and identify any shortcomings or deficiencies in the development of the ICAO global transition plan, in particular any portion which is more specifically relevant to that region, and to develop a plan for regional implementation aimed at ensuring the timely, efficient, orderly and coordinated transition to the new CNS/ATM systems. A plan for transition to CNS/ATM systems has since been developed describing regional characteristics along with proposed technical approaches in order to resolve the present shortcomings (ICAO, 1996).

North Atlantic Systems Planning Group (NAT SPG) The NAT SPG has been reviewing the impact of ADS on controller workload, the use of different protocols in pre-operational and engineering trials, as well as developing a NAT proposed ADS-ATS implementation plan, the certification of ATM systems of aircraft and operators and the NAT implementation programme. A NAT Implementation Management Group was established to oversee planning and implementation issues concerning CNS/ATM. A concept document, providing a blueprint for ATM improvements has been committed to (ICAO, 1996).

Canada Mexico United States CNS/ATM Implementation and Transition Plan In June, 1995, Canada, Mexico and the United States held their first CNS/ATM Working Group Meeting in Merida, Mexico. At that meeting, they agreed to pursue a CNS/ATM plan for North America. In reaching this agreement, the three States have committed to the incremental development of a plan which clearly identifies near-term, mid-term and long-term efforts and goals leading to a harmonized implementation of the new CNS/ATM systems on the North American continent. They agreed that their plan should be complementary to plans being developed by other ICAO regional planning groups (Canada, Mexico, United States, 1995). Additionally, the three States agreed that initial identification of current efforts and projects related to CNS/ATM within their region would not only be the basis of the CNS/ATM plan, but would also form the cornerstone of future planning.

The result of the meeting was the development of a draft CNS/ATM planning document consisting of an implementation and transition plan, based on CNS/ATM systems. Their implementation strategy identifies jointly agreed objectives and actions for implementation of system components to meet current and future user requirements and technological developments. The transition identifies priorities and a set time table for the integration of the various components into a fully harmonized Canada, Mexico, United States CNS/ATM system which will also be compatible with adjacent regional plans and with the ICAO Global Plan.

The ultimate goal of their initiative is to create a North American Regional Air Navigation Plan for submission to ICAO.

Free Flight

Free Flight is the guiding vision, mission and operational concept of the Federal Aviation Administration (FAA), based on CNS/ATM systems. At its basis, the concept will enable optimum flight paths for all airspace users through the application of CNS/ATM technologies and the establishment of air traffic management procedures that maximize flexibility while assuring positive separation of aircraft.

The boundaries and limitations of the concept continue to be explored, however, the concept has been defined as a contract-free environment to the airspace user, with user optimal flight paths and increased operational flexibility, while maintaining assured separation. The concept is based on two zones, *protected* and *alert*, the sizes of which are based on the aircraft's speed, performance and communication, navigation and surveillance equipment. The protected zone, the one closest to the aircraft, can never meet the protected zone of another aircraft. The alert zone extends well beyond the protected zone and, upon contact with another aircraft's alert zone, signals that intervention will be required. In principle, the alert zone should allow total flight path flexibility (RTCA, 1995).

The concept would allow pilots to operate their flights without being assigned specific routes, speeds or altitudes. Aircraft would only be restricted when there is a conflict with another aircraft, in particularly dense traffic such as near busy airports, in special use airspace, or when controllers issue limits due to safety of flight (Hughes, 1995).

Implementation Free Flight will be a continuum that ranges from total flight path flexibility to four dimensional contracts.

Dynamic density, a measure of traffic density and complexity of flow and separation standards, determines where a portion of airspace is on this continuum and the degree to which Free Flight can be allowed. It is expected that there will be dynamic swings between Free Flight and four dimensional contract usage. The challenge is to refine the definition of dynamic density and establish the values at which the yet to be established Traffic Flow Management Units, can most effectively switch between these two alternatives.

The early stages of Free Flight are being implemented in the United States' National Airspace System. The National Route Programme is conducting an experiment that allows users to both file and fly direct routings in the upper airspace of the National Airspace System. The ultimate goal of the experiment is to allow all aircraft above 31,000 feet to take part and to take advantage of Free Flight (FAA, 1995).

There are still many questions that must be addressed by States, professional organizations, researchers, etc, ranging from where authority for air traffic control will be, or can safely be placed, to what role should automation play and what the human's role should be. Some of these issues are addressed in Chapter 8. It is necessary that all of the various groups and stakeholders concerned take part in the debate.

The European Air Traffic Management System (EATMS)

In Europe, the planning and execution of ATM has been carried out mainly on a national basis with varying degrees of coordination via organizations such as Eurocontrol, ICAO and the European Civil Aviation Conference (ECAC). The public perception of air transport in Europe has been poor for quite some time, due mainly to the excessive delays that have become so commonplace in the system (Eurocontrol, EATMS Mission, Objectives and Strategy Document [MOSD], 1995b).

ECAC strategy for the 90s The extent of the problems and the certainty of further increases in European air traffic were recognized in the late 1980s. The Transport Ministers of ECAC States adopted the Enroute ECAC Strategy for the 1990s on 24 April 1990. To achieve its objectives, the European Air Traffic Control Harmonization and Integration Programme (EATCHIP) was structured in overlapping phases (Eurocontrol, 1995a) of which the last one, Phase IV, is aimed at the following:

* Adoption of a common functional model integrating the airborne and ground-based components of the future European Air Traffic Management System;

* Definition of transition programmes based on the common model;

* Implementation in specified zones of advanced systems supported by extensive automation and enhanced data communication available with Mode S, satellites and the aeronautical telecommunication network; and,

* Progressive extension of implementation of advanced systems to other zones.

In 1995, the Mission, Objectives and Strategy Document (MOSD) was developed as a top level policy document meant to identify the essential, basic principles which must be clearly kept in sight throughout the EATCHIP definition and implementation process.

The Introduction to the MOSD (1995) states that the elaboration of the document is founded on the principles formulated within the context of CNS/ATM systems, and that it will comply with the ICAO CNS/ATM plan for the European Region.

The mission statement of EATMS is to allow all airspace users the maximum freedom of movement subject to the needs for safety, cost-effectiveness, environmental aspects and national security requirements.

The Program for Harmonized Air Traffic Management Research in Europe (PHARE)

PHARE is an experimental program being sponsored by EUROCONTROL and national ATC authorities in the United Kingdom, France, Germany and the Netherlands. Canada and the United States are associate members in the project. The initial goals of the research have validated several new technologies and computer-assisted tools, and integration of airborne and ground-based systems was successfully demonstrated. The goal of PHARE is to develop computerized tools and data links to enable controllers to safely handle increasing amounts of air traffic. The aim is to minimize the amount of intervention required by controllers. Much like the Free Flight concept of the United States, aircraft crews would select a preferred flight path that would be transmitted to ground controllers who would intervene only if there was a conflict with other aircraft (Morrocco, 1996). Many PHARE program officials have stated their belief that the United States and Europe must have compatible systems and have therefore increased their links with the FAA. Of particular interest in the PHARE system, is an experimental flight management system (EFMS) that makes use of GPS and other distance measuring equipment to determine an aircraft's position. EFMS can then continuously calculate the demands required to keep the aircraft within the parameters of an ATC clearance in terms of time and space.

Africa

Africa is composed of many delicate national economies separated by vast wilderness areas. The costs of installing and maintaining a ground based infrastructure for the provision of air navigation services is prohibitive. The practical difficulties in maintaining ground navaids and VHF relay stations thousands of miles away from the major cities are insurmountable. As a given, the economic difficulties that much of the continent is now faced with, the solution to the air navigation problem must also be relatively inexpensive (Learmount, 1993).

Poor telecommunications adversely affect aircraft operations in many ways. Deficiencies in the aeronautical fixed telecommunication network (AFTN) negatively affect the exchange of messages between ATC units, leading to inadequate coordination of air traffic and allocation of flight levels and unavailability of the most fuel efficient routes. This problem also affects the dissemination of meteorological information.

High Frequency (HF) radio is the most common means of air to ground communications, but is certainly less than satisfactory. It has an extremely poor quality and limited amount of available frequencies.

In order to address the problems and improve overall efficiency of the airspace, the African Planning and Implementation Regional Planning Group (APIRG), in 1991, endorsed CNS/ATM systems and began development of implementation plans.

Recognizing the significance of aeronautical satellite communications to CNS/ATM, APIRG invited the International Maritime Satellite Organization (Inmarsat) to participate as an observer in its meetings and is looking to that organization to help arrange practical demonstrations of its technology at work (Learmount, 1993).

In the APIRG plan for the future, priority will go to aeronautical fixed and mobile communications via satellite. It is also envisaged that satellite navigation using GPS will take over for the ground systems now in place.

ICAO is playing a direct role in upgrading the air navigation infrastructure in the African Region by supplying technical expertise and education. Additionally an effort to acquaint pilots and controllers with the CNS/ATM plan is also being carried out mainly through seminars and workshops.

The African Region is an unusual one, offering unique opportunities. The Region could benefit from CNS technology more than most other continental areas in the world. The full benefits will come not only from the implementation of the newer technologies such as ADS, but from the adoption of an air navigation plan that is based on a required total system performance that includes ASM, ATS, ATFM and flight operations, based on the regional ATM concept. Such a plan would allow for a more flexible airspace, providing more dynamic routing possibilities and better overall efficiency and economies of flight. For this to happen, planning must be based on the identification of the ATM requirements for communication, navigation and surveillance and the incremental implementation of an integrated ATM system based on these requirements.

The above planning requires close coordination and cooperation at the regional level.

The FANS Stakeholders' Group (FSG)

The FANS Stakeholders' Group consists of organizations with an interest in implementing CNS/ATM and the capability to contribute constructively to implementation efforts. The FSGs goal is to ensure the timely coordinated and cost effective implementation of CNS/ATM systems on a global basis.

The FSG may provide a means through which resources can be dedicated to assist governments of ATS providers and airlines to identify impediments and initiate action to overcome obstacles in the implementation of CNS/ATM elements. The FSG attempts to build on the experience gained in one part of the world in order to assist in other regions, thereby reducing time and costs in implementing CNS/ATM.

The founding members of the FSG represent a cross section of the industry involved in implementation of CNS/ATM and form the Board of the FSG. The FSG brings together end users and organizations involved in the making of policy and technical standards, and companies that provide equipment and services. The founding members provide support for the FSG including resources and funding.

The international groups on the board are responsible for representing the interests of their individual members at board meetings. Private companies with the ability to contribute to implementation efforts, also participate in FSG activities. The management team and members of the FSG work closely with ICAO Headquarters and the regional offices in order to ensure that implementation of systems conforms to ICAO standards. Figure 6.6 shows an action programme for implementation of CNS/ATM which displays the responsibilities of the different stakeholders.

Finally

* The Air Navigation Commission has recognized that the operational requirements for ATM must first be established, followed by the necessary development to fulfil the operational needs. Based on this philosophy, the Commission is pursuing its work on ATM under a specific task aimed at development of operational requirements for global ATM;

* A structured approach to the work of ATM has been developed by the Air Navigation Commission which covers the ATM

ICAO	States/Regions	Service Providers (e.g. Inmarsat, ARINC)	Users (e.g. airlines)	Aviation industry
• Develops Standards and Recommended Practices (SARPs) including guidance material • Formed a project team • Formed an advisory group for monitoring and co-ordination • Continues to help States/regions develop plans in accordance with the global plan • Develops and assists in training • Provides technical assistance and assistance with cost/benefit analyses, including seminars as needed • Support continued allocation of spectrum to meet requirements • Update ICAO Global Plan • Develop an ATM operational concept • Develop institutional guidelines	• Follow ICAO implementation guidelines • Establish requirements • Develop and implement regional plans with ICAO regional offices • Perform cost/benefit analyses • Engage in research, development, trials and demonstrations (RDT&D) • Select implementation options • Co-ordinate with service providers • Co-ordinate with ICAO • Implement early applications • Adhere to institutional guidelines • Develop and conduct training • Adhere to procedures and practices • Remove obsolete equipment • Conduct certification including RNP	• Develop and install necessary infrastructure • Participate in standards development • Be involved in RDT&D • Co-operate with each other and with air traffic services	• Install avionics • Develop equipment standards • Co-operate in planning the transition • Be involved in RDT&D • Exploit applications – AMSS • ATS • AOC • AAC • APC – GNSS • En-route • Terminal • Non-precision approach • Precision approach • Participate in training	• Participate in standards development • Participate in RDT&D • Support transition planning activities • Assure the provision of adequate logistics support and training for new CNS equipment

Figure 6.6 Action programme for implementation of CNS/ATM which displays the responsibilities of the different stakeholders

requirements for communication, navigation and surveillance, as well as related aspects of airspace management, air traffic services and air traffic flow management required to support the evolutionary development of an integrated, global ATM system. Ten new Commission tasks have been established to pursue these goals.

Recapping the major points

A general objective of ATM is to enable aircraft operators to meet their planned times of departure and arrival and adhere to their preferred flight profiles with minimum constraints without compromising agreed levels of safety. The primary objective of air traffic control is centred on safety.

Some other objectives to be reached through the envisaged ATM system can be summarized as follows (ICAO, 1993a):

* To meet evolving air traffic demand;
* To support a safe and orderly growth of international civil aviation;
* To enhance safety, regularity and efficiency;
* To enhance economy of commercial air transport;
* To optimize benefits through global integration.

This chapter provided background on the work of the FANS Committee and ICAO as that work is related to the ATM component of CNS/ATM. This included a description of the changes that will form the basis of future systems and methods, and a review of the benefits to aviation that could be expected with implementation of these systems.

This review also included an analysis of work being undertaken toward the development of international standards, procedures and operational requirements associated with ATM and how these will form the underpinning of an integrated ATM system.

Before developing the standards necessary for harmonization and integration, an operational concept of the future ATM system first needs to be defined. This will clarify the benefits and give States and industry a clear objective for designing and implementing ATM systems.

As part of the operational concept, a total system performance is being defined. The total system can be seen as the totality of airspace, flight operations, and the facilities and services provided. When fully defined, RTSP will include criteria that should be met by the entire ATM system in the areas of safety, regularity, efficiency, sharing of airspace and in the area of human factors.

ATM consists of several subelements; these are: ASM, ATS and ATFM. In the future ATM system these subelements will evolve and take on different roles than exist today, mainly because they will integrate into a total ATM system that would also include flight operations.

The main scope of ASM will be toward a strategic planning function of airspace infrastructure, and flexibility of airspace use. As regards ATM, ASM is seen to consist of two main elements: first, the determination, for any given airspace, of the ATM requirements for communications, navigation and surveillance; and second, infrastructure planning.

The ATM requirements for communications will specify the air-ground and ground-ground communication coverage required for both voice and data link, supported by the technological developments outlined in Chapter 3 (e.g., satellites, data link) which would lead toward more effective ATM.

The ATM navigation requirements will specify a requirement for area navigation (RNAV) capability by all aircraft, supported by the technological developments outlined in Chapter 4 (e.g., GNSS, RNAV).

The ATM surveillance requirements will specify criteria as to where radar and/or ADS coverage is required for ATM purposes, supported by the technological advancements outlined in Chapter 5 (e.g., Mode S, ADS).

Optimum use of the airspace, meeting the efficiency objectives of future ATM will not be realized unless the airspace infrastructure is developed in a harmonized manner and in accordance with global efficiency requirements and safety levels. Future ATM therefore requires a cooperative approach. Airspace infrastructure planning should be applied over larger geographical areas than is currently the practice. These air traffic management regions should encompass homogeneous areas of adjacent flight information regions with compatible characteristics and requirements.

The functional integration of all of the elements of the ATM system is the key to its success.

In areas of high density traffic it will be necessary to implement powerful automation systems to respond to the ever increasing demand. The use of ATM automation tools will help air traffic controllers and the ATM system as a whole to sort, process and display information. This will assist air traffic managers to visualize the data and consider the range of alternatives.

Conclusions (recapping the major points)

As we move toward the future, there is likely to be controversy and debate as the various entities and stakeholders attempt to foresee what the effects of the new systems and operational concepts will be.

Captain Rob McInnis (1995), President of the International Federation of Airline Pilots Association (IFALPA), recently offered his very prudent and discerning views as to how the international civil aviation community should move toward and view the future as follows:

> The future of our aviation industry is built upon the efforts of the past. What we may see as revolutionary and completely different from our current way of doing things may be perceived as a simple evolutionary development by our inheritors. The difference is perspective. We should not close our minds to any ideas that will help to serve our community. We are obligated to investigate the new concepts. We have a responsibility to our professions and the public to ensure the safety of any system that is implemented. Furthermore, we are the ones who must advise the drivers of the aviation industry that the greatest of dreams is not easily achieved without hard work, determination and consistent effort to bring it into existence (p. 16)

Questions and exercises to expand your knowledge

1) What is the primary objective of air traffic control?

2) Explain how the elements of communication, navigation and surveillance, as envisaged in the future concept, will support ATM.

3) Describe the term functional integration as it relates to the future ATM system.

4) Describe the "total system" as explained in this chapter and as envisaged in the future ATM system.

5) When defined, the Required Total System Performance (RTSP) will become an important element in regional planning activities. How will this affect the regional planning processes and lead to an integrated ATM system.

6) When determined, ATM requirements for communication, navigation and surveillance will simplify the decision making process and also reduce the expense associated with aircraft equipage. Why is this so?

7) Describe how automation in the future ATM system will increase capacity of the system.

8) Imagine a flight management system in an aircraft communicating directly with ground computers, as part of a future ATM system. Identify some of the information that could be exchanged and describe a few of the advantages that this exchange would offer.

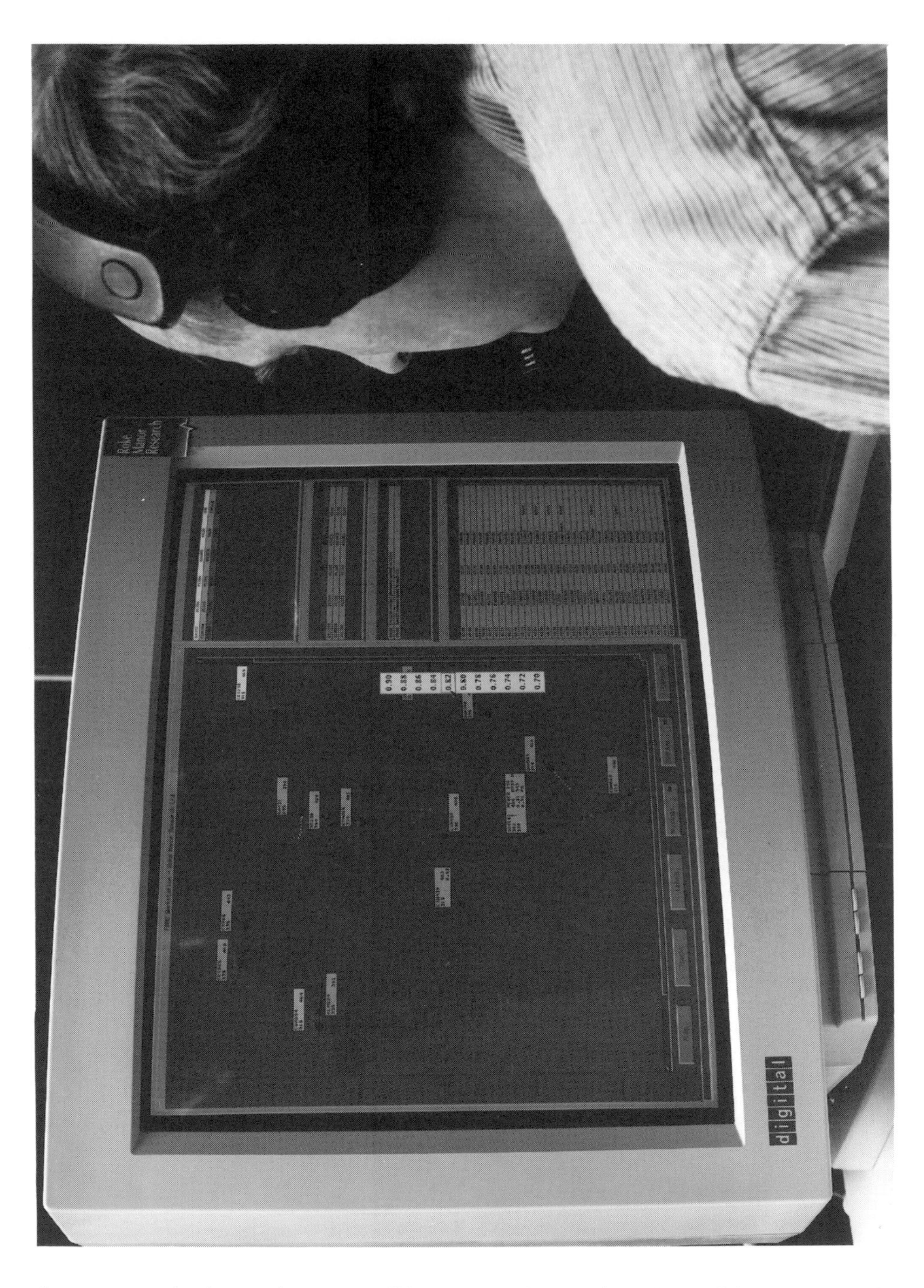

An automatic dependent surveillance (ADS) workstation with enhanced flight data blocks and associated flight data information to keep track of aircraft. (Picture provided courtesy of Roke Manor Research.)

7. Transition and implementation

Introduction

This chapter deals with CNS/ATM evolution and transition issues. It identifies critical transition events and addresses several key issues in terms of the human element, equipment and facilities, operations and costs involved. Training issues are reviewed and guidelines for transition to communications, navigation and surveillance systems are outlined. Finally, implementation options are listed for States and regional planning groups to follow. The need for coordination of planning activities at the global level is stressed in order to ensure a smooth transition to CNS/ATM systems. Institutional issues are briefly addressed; however, this is a particularly difficult issue and requires a great deal of study and effort that is beyond the scope of this text.

The chapter is a partial reproduction of Sections Four: Global Planning, and Five: System Implementation, from the ICAO document: *The ICAO CNS/ATM Systems: Coping with Air Traffic Demand*, which is found at Appendix B to the Report on Agenda Item 8 of the Fourth Meeting of the Special Committee for the Monitoring and Coordination of Development and Transition Planning for the Future Air Navigation System (FANS II) (ICAO, 1993a).

The chapter makes reference to the Global Plan, a document that was produced by the FANS (II) Committee. That document is presently (1996-97) being updated by the ICAO Air Navigation Bureau under the guidance of the ICAO Council and the Air Navigation Commission to more realistically reflect the changes that have already taken place, to give practical guidance to States and regional planning groups and to adapt to changing technologies. This will take into account work being done at the regional and worldwide levels as well as by individual States.

Global planning

Global planning will be essential to the successful implementation of the new CNS/ATM systems. The global plan sets forth the details of this planning. The discussion in this section provides an overview of the key issues involved.

Evolution and transition

The need for an evolutionary transition is critical to planning for the new CNS/ATM systems. The transition should be carefully planned to avoid degradation in system performance. The level of safety attainable today will need to be assured throughout the transition. Careful planning will also be necessary to ensure that aircraft of the future are not unnecessarily burdened by the need to carry a multiplicity of existing and new CNS equipment during a long transition cycle.

There is a close relationship between the required CNS services and the desired level of ATM. For reasons of both economy and efficiency, it is necessary to ensure that differences in the pace of development around the

On a global scale, the transition to the ICAO CNS/ATM systems will be paced by a series of key events that must be completed for the implementation to proceed. Activities relating to those events are currently being pursued under the sponsorship of ICAO and/or its Contracting States. Global transition planning requires recognition and understanding of those events and the associated schedule of activities so that the individual planning undertaken by ICAO, States and international organizations can lead to the intended benefits. There are many such events, but for planning purposes it is useful to group them as follows:

- The availability of standards and procedures
- The completion of necessary trials and demonstrations
- The availability of adequate satellite capacity (when applicable)
- The equipage of suitable numbers of aircraft
- Operational use
- Training

Figure 7.1 Key transition events

world do not lead to incompatibility among elements of the over-all system. In particular, given the wide coverage of satellite CNS systems, world-wide co-ordination of these systems is necessary if they are to be optimized.

It is recognized that there are major long-term consequences of adopting a CNS concept that will eventually permit the elimination of a variety of current CNS systems. Decisions on whether particular systems can be removed will depend on many factors. One essential factor is the demonstrated capability and implementation of the new systems. Moreover, a clear and compelling case for transition to the new CNS system will include consideration of the benefits perceived by the aviation community.

General transition issues

Because of differences in the level of ATM in various parts of the world and the variety of other factors influencing the transition, exact time frames for the transition cannot be specified.

Ideally, the transition to new CNS systems should be based on improvements in ATM, and should be accompanied by structural and procedural changes that will enhance ATM and provide benefits to users. The necessary structural changes involve airspace re-organization required to optimize the new systems. The necessary procedural changes include:

* Data link handling procedures;
* Review of ATC procedures, including separation criteria;
* Review of separation minima;
* Message formats;
* Approach procedures.

Planning and implementation of improved ATM capabilities should include consideration of human factors impacts and requirements. Essential human factors considerations for the transition include:

* Training (discussed below);
* Human/machine interface;
* Definition of safety level with reference to system statistics as well as human capabilities and limitations;
* The retention of situational awareness by ATS personnel and air crews;
* Provision for user-preferred routings;

Human

- Procedures and training for use of parallel systems
- Human/machine interface issues for parallel systems
- Operator/user confidence in the new system
- Selection criteria for operators/users of the new system
- Procedures in case of new or old system failure
- Automation issues
- Operator knowledge of the system mix

Equipment/Facilities/Infrastructure

- Availability of equipment, facilities, and infrastructure for new and old
- systems
- Maintainability of new and old equipment, facilities, and infrastructure types
- Partial or non-standard equipage within a facility or aircraft
- Partial equipage across facilities or an aircraft fleet
- Integrity assurance and back-up for new and old systems
- Multiple equipment types or facilities for the same function
- Design of new systems to allow for further upgrade
- Replacement of aging components of the old system
- Upgrading of aircraft/facilities with limited remaining life
- Capacity and coverage issues

Certification of equipment for the new system

Operational

- Use of different systems in different flight phases
- Different aircraft capabilities or user classes in the same airspace
- National or other boundaries with different infrastructures
- Non-equipped "intruders"
- Disruption of operations to change the system
- Supplementary versus sole-means use
- Aircraft/facilities out of service to re-equip
- Initial benefits of the new system during the transition

Management

- Standards and regulations
- Phased transition; intermediate steps
- Organizational restructuring during the transition
- Pre-operational trials
- Research and development and applications development
- Need and incentives to minimize the duration of the transition period
- Communication among States on implementation plans

Cost

- New system cost and investment in the old system; amortization
- Cost of maintaining parallel systems

Figure 7.2 General transition issues

* Effective packaging and managing of information relevant to users and ATS personnel;
* Definition of responsibilities of pilots, air traffic controllers and system designers in an automated environment.

Training considerations

Training requirements associated with new technologies are an essential concern, especially for less developed States. In those States, a lack of training is an important reason given for the inability to implement existing systems. Moreover, training considerations associated with the new ICAO CNS/ATM systems may include the need to identify new ratings for controllers, and would thus require amending Annex 1.

Training is generally considered the responsibility of each State. However, it is understood that many States will need guidance in this area, and perhaps assistance.

It should be noted that implementation of the ICAO concept may result in a net reduction in training requirements for CNS elements over the longer term. To explore this possibility, the FANS Committee addressed training issues in the following three categories:

Areas of training that would no longer be required as a result of implementing the ICAO system elements

With existing CNS systems, there are noticeable training deficiencies at the technical level in many areas of the world. These can result in excessive system down-time and sub-standard system operation. As a result of implementing the ICAO CNS/ATM elements, many of the current systems will be progressively withdrawn, and with them, the need for training in those systems.

New areas of training required by the ICAO system elements

The training required at the technical level will depend on the specific system implementation, and the extent to which systems such as ground earth terminals are provided by the ATS authority, rather than having services provided via contract. Thus, comprehensive training will be required at the managerial level to address the new and often complex institutional, administrative, and economic issues associated with the transition to the new system including the aeronautical telecommunication network (ATN).

Implementation of the new CNS/ATM systems will result in the automation of many ATC functions previously performed manually by air traffic controllers. As a result, interaction between controllers and flight crews will take on different dimensions. Thus, training of both controllers and flight crews must take into account the full implications of the automation, including back-up procedures to be used in the event of system malfunction.

Possible sources of training support

Sources of financial and technical support for required training will need to be developed. Possible sources for such support include the United Nations Development Programme (UNDP), civil aviation training centres (CATCs), the TRAINAIR Central Unit of ICAO and State sponsored in-house training programmes.

Under the United Nations Development Fund (UNDP), the execution of planned programmes and projects is delegated to international organizations as executing agencies. In the case of aviation projects, ICAO serves in this capacity. The degree of ICAO's participation is determined by the individual requests submitted by governments of developing States, which are responsible for deciding what portion of the total assistance made available to them by the UNDP should be used for civil aviation.

Over seventy civil aviation training centres have been established throughout the world. These CATCs provide the majority of trained personnel operating national and regional air transport systems.

Many universities around the world are increasingly developing programmes that aim at placing professional aviation personnel in the industry. Universities will certainly have an important role to play in research, development and training for CNS/ATM.

Contracting States in a position to do so could provide direct training support to States needing or requesting CNS/ATM training. Such Contracting States could send technical experts to the requesting States to provide in-house training. A State receiving training would enter into a suitable reimbursement arrangement with the State providing the training. Additionally, several States have training facilities that could be used to provide CNS/ATM training to other States. These types of direct training programmes could be coordinated through ICAO to ensure that the training provided by possibly more than one State is standardized and consistent in quality.

Relationship to other ICAO planning activities

Detailed transition and implementation planning continues to be a regional responsibility. It is essential to maintain and improve co-ordination among the global and regional planning activities through which the global plan for the new system is to be implemented. ICAO has determined that regional planning groups are a better mechanism for this purpose than the traditional regional air navigation meetings, which tend to be held infrequently. The role of regional planning groups is of paramount importance in providing impetus for implementation of the global plan and ensuring its regional co-ordination throughout the various ICAO regions.

The need for co-ordination can be considered in terms of time phases, with somewhat different requirements for action in each phase. Initially, co-ordination among those regions involved in planning will be informal, and based on mutual awareness between ICAO and the regional planning groups. In the mid-term, more formal action, based on the mutual co-ordination of planning documents, should be carried out. Finally, in the longer term, there will be a continuing need for co-ordination. ICAO will develop the mechanisms for this long-term task as events develop.

System implementation

The action programme for transition to and implementation of the new CNS/ATM systems includes guidelines and key activities for the States/regions, users, service providers, and ICAO.

The action programme outlined in this section is intended to provide users, service providers, and the States/regions with guidelines for transition to and implementation of the new CNS/ATM systems.

Transition guidelines

Guidelines for transition to the future system encourages early equipage by users for the earliest possible accrual of the system benefits. Although a transition period of dual equipage, both airborne and ground, will be necessary to ensure the reliability and availability of the new system, the guidelines are aimed at minimizing this period to the extent practicable.

Institutional considerations

ICAO is responsible for developing the position of international civil aviation in all matters related to the use of space technology for air navigation purposes. In accordance with this responsibility, the ICAO Legal Committee was tasked to study the FANS concept and report on any possible impediments. The committee concluded that FANS poses no threat to the sovereignty of States and is in compliance with the ICAO Convention.

- States should begin to use data link systems as soon as possible after they become available.
- Transition to the aeronautical mobile-satellite service (AMSS) should initially be in oceanic airspace and in continental en-route airspace with low-density traffic.
- States/regions should co-ordinate to ensure that where ATC applications supported by AMSS are to be introduced, they should be introduced approximately simultaneously in adjacent flight information regions (FIRs) through which there are major traffic flows.
- During the transition period after AMSS is introduced, the current levels of integrity, reliability, and availability of existing HF communications systems must be maintained.
- Communications networks between ATC facilities within a State and ATC facilities in adjacent States should be established if they do not already exist.
- The aeronautical telecommunication network (ATN) should be implemented in phases.
- If new application message processors and data link systems are implemented, they should support code- and byte-independent data transmission and be fully compatible with the ATN.
- During the transition, States should co-operate on a bilateral and multilateral basis to ensure operation of the ATN meets the needs of international aviation and the States.
- States should establish procedures to ensure that both security and interoperability aspects of the ATN are not compromised.

Figure 7.3 Guidelines for transition to the new communications system

- The global navigation satellite system (GNSS) should be permitted for supplemental en-route use first, and later for use as a sole-means system for en-route radio navigation.
- The ground infrastructure for current mandated navigation systems must remain available during the transition period.
- States/regions should consider segregating traffic according to navigation capability and granting preferred routes to aircraft with more accurate capability.
- States/regions should co-ordinate to ensure that separation standards and procedures for appropriately equipped aircraft are introduced approximately simultaneously in each FIR through which major traffic passes.

Figure 7.4 Guidelines for transition to the new navigation system

- States should begin to develop operational procedures, in accordance with ICAO Standards and Recommended Practices (SARPs), procedures, and guidelines, for the implementation of automatic dependent surveillance (ADS) within airspace under their control.
- Transition to ADS should initially be in oceanic airspace and in continental en-route airspace with low density traffic.
- States/regions should co-ordinate to ensure that where ADS is introduced, it is introduced approximately simultaneously in each FIR where major traffic flows occur.
- Where differing surveillance methods are employed in adjacent States or FIRs, commonality and comparability of systems are essential. Procedural means should be developed to make the service transparent to users.
- During the transition period after ADS is introduced, levels of integrity, reliability, and availability of existing position reporting systems must be maintained.
- States/regions should take actions within the ICAO framework to ensure that the implementation of procedural changes due to ADS and other systems will result in more efficient use of airspace.
- During the transition to ADS, suitability equipped aircraft should be given precedence over non-ADS-equipped aircraft for preferred routes and airspace.
- ADS should be introduced in phases.
- ADS equipment, standards, and procedures should be developed in such a way as to permit the use of ADS as a back-up for other surveillance methods.

Figure 7.5 Guidelines for transition to the new surveillance system

Institutional guidelines

To ensure that States and regions maintain the institutional capability of implementing the new CNS/ATM systems, the FANS Committee has developed institutional guidelines. The guidelines have been endorsed by the ICAO Legal Committee and appear in the global plan. They are intended to assist States and regional planning groups to assess the adequacy of proposals for the delivery of AMSS, ATN and GNSS services. The guidelines in the global plan are in four sections, covering:

* General considerations applicable to all systems;
* Considerations specifically applicable to AMSS;
* Considerations specifically applicable to ATN; and
* Considerations specifically applicable to GNSS.

Costs and charges

For early GNSS services, GPS and GLONASS satellite signals will be provided with no direct user charges for the foreseeable future. As the system matures, additional or replacement satellites, transponders on existing satellites, and ground equipment, if necessary, to provide integrity and other augmentations, will need to be implemented by States or groups of States or other organizations. As is also true for satellite communications, costs for such augmentations are expected to be recovered through charges to the users of the services.

States have an important degree of choice as to how the costs are incurred, e.g., in the form of capital expenditures, expenditures on leasing fees, or on a usage-based operating cost basis. States, or groups of States, may act in concert to ensure the provision of necessary services and share the costs. The use of communications or navigation service providers, in the public or private sector, may be a choice worthy of consideration. In all cases, however, significant costs for the provision and maintenance of current ground-based systems will be avoided.

Current cost recovery systems, where aircraft operators are charged for the services provided when passing through a State's flight information regions (FIRs), are not expected to change. ATS providers could also work jointly with adjacent States, as is currently the case in many areas, and cost recovery systems could be revised accordingly. One State could collect on behalf of another, or a multi-State body could be established with a revenue role, or ICAO itself might play a coordinating or administrative role in cost sharing arrangements. All such methods are in current use today.

Implementation options

States should take a logical approach in assessing implementation options for the new CNS/ATM systems:

* The initial phase should be to select a suitable implementation option. During this phase, States should establish their future requirements for CNS and ATM. Additionally, they should determine the implications for existing facilities/systems and undertake cost/benefit analyses. If the cost/benefit analyses turn out to be favourable for the chosen implementation option, action should be taken to put that option into effect; if not, other options should be considered through an iterative process. A Cost/Benefit Analysis Guide and an ICAO circular on cost/benefit and cost-effectiveness analysis are both available from ICAO to assist States with their studies;

* To initiate the process and assure the necessary participation of all parties, there are roles to be played by ICAO, States and regions, State service providers, and users;

* Satellites which will provide communication services in the AMSS band are already in existence and additional satellites may be expected in future years to meet increasing demand. Communication services furnished via such satellites are presently provided, and are expected to continue to be provided through entities called service providers. In order to evaluate and implement these services, the State providing air traffic services should contact service providers and reach an agreement on system performance parameters. Next, the State should consider arrangements for the provision of equipment and/or services, cost financing, training, cost recovery, certification, and ongoing service support. If the State accepts all proposed provisions, it should enter into contract negotiations for contract agreement and signing. From this point, a phased implementation programme

should go into effect, proceeding from the procurement of equipment and/or services through commissioning to operation;

* Full GNSS services are not currently being provided, although, in the very near future partial services will commence. Thus, for GNSS, the available options will depend on the type of services to be implemented and the RNP airspace in which they are to be provided (enroute, terminal, non-precision approach, etc.). ICAO has confirmed the need for the transition to GNSS to begin in the near term, using existing satellite capabilities, augmented as necessary to assure service integrity and availability for the particular RNP airspace. Accordingly, the implementation of GNSS will be accomplished through a phased implementation strategy. To achieve the phased implementation, several options are available, and the selection of a particular option will require a cost/benefit analysis. The options range from the use of existing satellites (GPS/GLONASS) augmented for integrity and availability assurance, to a "regional" option in which additional satellite constellations provided by other States or organizations are added to the existing satellites, to a longer-term GNSS, through the provision of new satellites by an international body, as yet undefined.

PART C

THE HUMAN ELEMENT

These tower cab integrated workstations display flight data on a high resolution monitor, allowing the controller to keep crucial data in constant view. This display takes the place of conventional paper flight strips. (Picture provided courtesy of Hughes.)

8. Human factors in CNS/ATM systems

Introduction

In 1986, the ICAO Assembly adopted Resolution A26-9 on Flight Safety and Human Factors. As a follow-up to the Resolution, the Air Navigation Commission formulated the following objective for the task:

> To improve safety in aviation by making States more aware and responsive to the importance of human factors in civil aviation operations through the provision of practical human factors material and measures developed on the basis of experience in States (ICAO, 1989, p. i).

One of the methods chosen to implement Assembly Resolution A26-9 was through the publication of a series of digests which would address various aspects of human factors and their impact on flight safety. The remainder of this chapter is comprised of excerpts of various sections of ICAO Digest No. 11, Human Factors in CNS/ATM Systems (ICAO, 1994i). Several issues concerning the role of the human controller in the future, highly automated CNS/ATM system are identified and addressed. The chapter deals with many issues related to automation in CNS/ATM systems and points to potential problems, based on past experience in other, safety critical industries, and identifies possible solutions.

ICAO Digest No. 11, on Human Factors in CNS/ATM systems, was produced by the ICAO Secretariat with the assistance of the ICAO Flight Safety and Human Factors Study Group. It is based mainly on the work of

Dr. Charles E. Billings, formerly of the Ames Research Centre, NASA, Moffett Field, California (1991) on the subject of human-centred aircraft automation. It has also borrowed considerably from Harold E. Price's "Conceptual System Design and the Human Role" (1990), published in MANPRINT — *An Approach to System Integration*, edited by Harold R. Booher, Van Nostrand Reinhold, New York, (1990). The International Federation of Air Traffic Controllers (IFATCA) is also recognized for its detailed review of and feedback on the original draft.

Other chapters in that digest have been left out as they are considered beyond the scope of this text. Digest No.11 is available through ICAO and is recommended reading for anyone with a serious interest in human factors in future aviation systems.

Historical background

The Tenth Air Navigation Conference (ICAO, 1991a) recognized the importance of human factors in the design and transition of future ATC systems. It also noted that automation was considered to offer great potential in reducing human error. It further recommended that work conducted by ICAO in the field of human factors pursuant to ICAO Assembly Resolution A26-9 include studies related to the use and transition to future CNS/ATM systems.

Following the recommendation of the Tenth Air Navigation Conference, the ICAO Air Navigation Commission agreed that its task: *Flight Safety and Human Factors,* should include work on human factors considerations in future aviation systems with an emphasis on CNS/ATM-related human-machine interface aspects.

Based on the decision of the Air Navigation Commission, the ICAO Secretariat contacted experts from selected States and international organizations and reviewed recent and ongoing studies to identify human factors issues of relevance to ICAO CNS/ATM systems. The survey identified several areas in which application of human factors knowledge and experience would enhance future ICAO CNS/ATM systems safety and efficiency as follows:

* *Automation and advanced technology in future ATM systems* The application of state-of-the-art technology and automation is fundamental to CNS/ATM systems. Experience shows that it is

essential to take into account the human element during the design phase so that the resulting system capitalizes upon the relative strengths of humans and computer-based technology. This approach is referred to as a "human-centred" automation.

* *Flight deck/ATM integration* ICAO CNS/ATM systems will provide for a high level of integration between aircraft and the air traffic control system. This will bring new and different challenges. The various components of the system will interact in new ways, and new means of communication between pilots and air traffic controllers will be available. A dedicated systems approach must be adopted to address the issues associated with this integration and to ensure that the system as a whole is user-friendly.

* *Human performance in future ATM* The human element is the key to the successful implementation of CNS/ATM systems. A broad base of scientific knowledge of human performance in complex systems is available and research continues to provide more. Additional research is still needed regarding the influence of organizational and management factors on individual and team performance in ATM. Information transfer in complex systems, the system-wide implications of datalink implementation, automated aids such as conflict prediction and resolution advisory systems, and the allocation of authority and functions between air and ground in future systems are areas in which guidance is necessary.

* *Training, selection and licensing of controllers* Acquiring technical skills alone will not guarantee on-the-job performance with high reliability and efficiency. Resource management training programmes specially tailored to ATM requirements are under development. Although some early attempts to address human factors training for controllers are in place, it is evident that much is lacking and more action in this regard is still desirable. Selection criteria which go beyond consideration of the candidate's technical aptitude and include social and personal characteristics associated with team performance are also important issues which are at the development stage. Licensing requirements which reflect these new training objectives would provide the framework to achieve them.

* *Safety monitoring of ATM activities* Existing tools for monitoring safety may not be sufficient in view of the increased complexity and interdependence of the ICAO CNS/ATM activities. Guidance is needed on how ATS activities can be monitored to provide the information required for identifying and resolving safety issues.

The material reproduced in this chapter addresses the first of these issues utilizing experience gained from human factors knowledge. It presents the human factors implications of automation and advanced technology in future aviation systems, including CNS/ATM systems. The remaining issues continue to be addressed through research, studies and at international fora. Much analysis and work remains to be done. ICAO will make information available as it is developed. Information can also be obtained from national aviation authorities.

Additional sources of information include ICAO Human Factors Digest No. 5 — *Operational Implications of Automation in Advanced Technology Flight Decks*, Human Factors Digest No. 8 — *Human Factors in Air Traffic Control* and ICAO Doc9583 — *Report of the Tenth Air Navigation Conference*.

Automation in future aviation systems

One major issue in future aviation systems, including the CNS/ATM system, is the impact of automation and the application of advanced technology on the human operator. In order to be effective, automation must meet the needs and constraints of designers, purchasers (i.e. civil aviation authorities) and users. It is, therefore, essential to provide guidelines for the design and use of automation in highly advanced technology systems including the CNS/ATM system. What roles should automation play in future systems, how much authority should it have, how will it interact with the human operator and what role should be reserved for the human are but a few of the many questions that are now being advanced and should be answered during conceptual system design.

The role of the human operator in highly automated systems

Technology has advanced to an extent for computers (automation) to be able to perform nearly all of the continuous air traffic control and surveillance as well as aircraft navigational tasks of the aviation system. Why, then, is the human needed in such systems? Couldn't automation be constructed to accomplish all the discrete tasks of the human operator? Would it not be easier and even cheaper to design highly reliable automata that could do the entire job without worrying about accommodating a human operator?

Many system designers view humans as unreliable and inefficient and think that they should be eliminated from the system. (This viewpoint is fuelled by the promise of artificial intelligence and recently introduced advanced automation.) It is unrealistic to think that machine functioning will entirely replace human functioning. Automation is almost always introduced with the expectation of reducing human error and workload, but what frequently happens is that the potential for error is simply relocated. More often than not, automation does not replace people in systems; rather, it places the person in a different and, in many cases, more demanding role.

The aviation system consists of many variables that are highly dynamic and not fully predictable. Real-time responses to developing situations are what assure the safe operation of the whole aviation system. The basic difference in the way humans and computers respond to situations could mean the difference between a reliable (safe) and an unreliable (unsafe) aviation system. Human response involves the use and coordination of eyes, ears and speech and the ability to respond to unexpected problems through initiative and common sense. Automation (computers) rely on the *right programme* being installed to ensure that the *right action* is taken at the *right time*. The inability of automation designers to engineer a programme that can deal with all presumed eventualities and situations in the aviation system, and the uncontrollable variability of the environment are some of the major difficulties of computerizing all the tasks of the aviation system. The reality is: if automation is faced with a situation it is not programmed to handle, it fails. Automation can also fail in unpredictable ways. Minor system or procedural anomalies can cause unexpected effects that must be resolved in real time, as in the air traffic control breakdown in Atlanta, Georgia (U.S.A.) terminal airspace in 1980 and the breakdown of telecommunication systems in New York in 1991. Considering these limitations, it is not very difficult to see that an automation-centred aviation system can easily spell disaster to the whole aviation infrastructure.

Although humans are far from being perfect sensors, decision-makers and controllers, they possess several invaluable attributes, the most significant of which are: their ability to reason effectively in the face of uncertainty and their capacity for abstraction and conceptual analysis of a problem. Faced with a new situation, humans, unlike automatons, do not just fail; they cope with the situation and are capable of solving it successfully. Humans thus provide to the aviation system a degree of flexibility that cannot now and may never be attained by computational systems. Humans are *intelligent:* they possess the ability to respond quickly and successfully to new situations. Computers, the dominant automatons of the ATC system, cannot do this except in narrowly defined, well understood domains and situations.

Automation should be considered to be a tool or resource, a device, system or method which enables the human to accomplish some task that might otherwise be difficult or impossible, or which the human can direct to carry out more or less independently a task that would otherwise require increased human attention or effort. The word "tool" does not preclude the possibility that the tool may have some degree of intelligence — some capacity to learn and then to proceed independently to accomplish a task. Automation is simply one of many resources available to the human operator, who retains the responsibility for management and direction of the overall system. This line of thinking has been well understood and precisely defined by the aviation human factors community, to the extent that philosophies have been developed by some organizations in the industry to demarcate the function and responsibilities of the two elements (human operators and automation) in the system. A very good example of such a philosophy as adopted by one operator states:

* The word "automation", where it appears in this statement, shall mean the replacement of a human function, either manual or cognitive, with a machine function. This definition applies to all levels of automation in all airplanes flown by this airline. The purpose of automation is to aid the pilot in doing his or her job;

* The pilot is the most complex, capable and flexible component of the air transport system, and as such is best suited to determine the optional use of resources in any given situation;

* Pilots must be proficient in operating their airplanes in all levels of automation. They must be knowledgeable in the selection of the appropriate degree of automation, and must have the skills needed to move from one level of automation to another;

* Automation should be used at the level most appropriate to enhance the priorities of Safety, Passenger Comfort, Public Relations, Schedule and Economy, as stated in the Flight Operations Policy Manual;

* In order to achieve the above priorities, all Delta Air Lines training programmes, training devices, procedures, checklists, aircraft and equipment acquisitions, manuals, quality control programmes, standardization, supporting documents and the day-to-day operation of Delta aircraft shall be in accordance with this statement of philosophy.

Introducing such an automation philosophy into aviation operations is beneficial since by defining how and when automation is to be used, it demarcates the boundary of human-machine responsibilities and thus promotes safety and efficiency in the system.

CNS/ATM system automation

The core of the benefits of the CNS/ATM system will be derived from automation intended to reduce or eliminate constraints imposed on the system. Data bases describing current and projected levels of demand and capacity resources, and sophisticated automated models that accurately predict congestion and delay will, in the future, be used to formulate effective real-time strategies for coping with excess demand. Automation will play a central role in establishing negotiation processes between the aircraft flight management computer systems and the ground-based air traffic management process, to define a new trajectory that best meets the user's objective and satisfies ATM constraints. The human operator, however, should decide the outcome of the negotiation and its implementation. Similarly, when the ground-based management process recognizes a need to intervene in the cleared flight path of an aircraft, the ATM computer will negotiate with the flight management computer to determine a modification meeting ATM constraints with the least disruption to the user's preferred trajectory. Automation can also probe each ADS position-and-intent report from an aircraft to detect potential conflicts with other aircraft, with hazardous weather or with restricted airspace.

The range of use of automated systems and automation is so central to the CNS/ATM systems that it will not be possible to derive the envisaged benefits of the CNS/ATM system or even implement it effectively without the use of automation. It is clear that the possibilities being researched as a result of the introduction of the global CNS/ATM system range well beyond what is strictly envisaged at present and further development may strictly depend on more and more automation.

Automation has been gradually introduced in the aviation system. Flight deck automation has made aircraft operations safer and more efficient by ensuring more precise flight manoeuvres, providing display flexibility, and optimizing cockpit space. Many modern ATC systems include automated functions, for example in data gathering and processing, which are fully automated with no direct human intervention. Computerized data bases and electronic data displays have enhanced data exchange, the introduction of colour radar systems have allowed a greater measure of control and the computerization of Air Traffic Flow Management (ATFM) has proved to be an essential element to efficiently deal with the various flow control rates and increases in traffic demand.

For the purpose of the ICAO Digest on Human Factors in CNS/ATM systems, automation refers to a system or method in which many of the processes of production are automatically performed or controlled by self operating machines, electronic devices, etc. The concern is with automation of future aviation-related technology and in particular with human factors issues in CNS/ATM systems development and application. Automation is essential to the progressive evolution of the CNS/ATM systems and is expected to play a commanding role in future development of aviation technology. As such, its progressive introduction, therefore, is most desirable.

The techniques of air traffic management are constantly changing. New data link and satellite communication methods are evolving, the quality of radar and data processing is improving, collision avoidance systems are being developed, direct routing of aircraft between departure and arrival airports instead of via airways is being explored, and future air navigation systems are being researched and developed. More and more possibilities intended to increase the benefits of the concept in a wider scale are also being discovered and introduced.

Further options offered by such technological advances have to be considered in terms of safety, efficiency, cost effectiveness and compatibility with human capabilities and limitations. These advances change the procedures and practices of the global aviation system, the working environment and the role of pilots, air traffic controllers, dispatchers, aircraft maintenance technicians, etc., presenting all involved with the challenge not to overlook the human factors issues involved. Whenever significant changes to operational procedures or regulations are contemplated, a system safety analysis must be conducted. The objective of such analysis is to identify any safety deficiencies in the proposed changes before they are implemented, and to ensure that the new procedures are *error tolerant* so that the consequences of human or technological failure are not catastrophic. Human factors consideration in the design and development of new systems can assure that the paramount requirement of safety is never compromised in the whole system, but maintained and enhanced throughout all future challenges.

Development in CNS/ATM systems will seek to do more with less, by designing and procuring air traffic management systems that are highly automated. Increased automation in aviation is inevitable. The issue is therefore about *when, where* and *how* automation should be implemented, not *if* it should be introduced. Properly used and employed, automation is a great asset. It can aid efficiency, improve safety, help to prevent errors and increase reliability. The task is to ensure that this potential is realized by matching automated aids with human capabilities and by mutual adaptation of human and machine to take full advantage of the relative strengths of each. In aviation automated systems, the human (pilot, controller, etc.), who is charged with the ultimate responsibility for the safe operation of the system must remain the key element of the system: automation or the machine must assist the human to achieve the overall objective, never the contrary.

A major design challenge in the development of air traffic management procedures and techniques using new technologies is to realize system improvements that are centred on the human operator. Information provided to the human operator and the tasks assigned must be consistent with the human's management and control responsibilities as well as the innate characteristics and capabilities of human beings. Any future technological advance in the aviation system, including the CNS/ATM system, should therefore take into account the human-machine relationship early in its design process and development. If account is not taken at this stage, the system may not be used as intended, prejudicing the efficiency or safety of the whole system. Automation must be designed to assist and augment the capabilities of the human managers; it should, as much as possible, be *human-centred.* As basic understanding of human factors improves, and as facilities for testing the human factors aspects of system designs become available, the design process can be expected to be easier.

Issues and concerns in CNS/ATM systems automation

CNS/ATM systems are intended to be a worldwide evolution of communications, navigation and surveillance techniques into a largely satellite-based system. As such, they entail a continuous increase of the level of automation in aviation operations. Optimum use of automation both in the aircraft and on the ground (air traffic control, dispatch and maintenance) is desired to permit high efficiency information flow. The Automatic Dependent Surveillance data will be used by the automated air traffic management system to present a traffic display with as much information as possible to the operator. To increase capacity and reduce congestion, airports and airspaces must be treated as an integrated system resource, with optimal interaction between system elements, aircraft, the ground infrastructure, and most importantly, the human operators of the system.

In some States, extensive research is being done on improvements to air safety through the introduction of air-ground data links replacing the majority of pilot/controller voice communications. It should, however, be recognized that voice communication will still be required, at least for emergency and non-routine communications. Automation is considered to offer great potential in reducing human error while providing for increased airspace capacity to accommodate future growth in air traffic. This, however, could involve changes in the human-machine interface which in the future may include increased use of artificial intelligence to assist the pilot and the controller in the decision-making process.

All forms of an automated assistance for the human operator must be highly reliable, but this may also induce complacency. Human expertise may gradually be lost and if the machine fails, the human operator may accept an inappropriate solution or become unable to formulate a satisfactory alternative. The most appropriate forms of human-machine relationship depend on the type of task which is automated and particularly on the interaction between planning and executive functions.

In the air traffic management environment, it is highly accepted that the performance of routine ATC tasks aids memory, which is not the case if these tasks are done automatically for the controller. Recent studies have shown that, in order to form a mental picture of the traffic situation, controllers derive a lot of their situational awareness by speaking to the aircraft and by making annotations on paper strips or making inputs (in more automated systems). Verbal and written (or keyboard) inputs keep people "in the loop" and allow active updating of the mental picture and situational awareness in its widest sense. It is believed that the automation of data can lead to deficiencies in human performance, since it can deprive the controller of important information about the reliability and durability of information. Automation may well reduce the effort required to perform certain tasks and the stress associated with them, but may also lead to loss of job satisfaction by taking away some of the intrinsic interests of the job, and the perceived control over certain functions.

There is enough information, both from safety deficiencies information systems and from accident reports, to illustrate the impact of the technology-centred approach to automation. More than 60 concerns relating to automation were identified by a subcommittee of the Human Behaviour Technology Committee established by the Society of Automotive Engineers (SAE) to consider flight deck automation in 1985. These concerns were grouped into nine categories, the majority of which are as relevant to the air traffic control environment as they are to the flight deck. A brief presentation of such concerns includes:

* *Loss of systems awareness* may occur when the human operator is unaware of the basic capabilities and limitations of automated systems, or develops erroneous ideas of how systems perform in particular situations.

* *Poor interface design* Automation changes what is transmitted through the human-machine interface, either leading to some information not being transmitted at all or the format of the transmitted information being changed. Traditionally, most information has been conveyed from the machine to the human by means of visual displays and from the human to the machine by means of input devices and controls. Poor interface design may also combine with the time required for the human to take over from automation and may become an important factor, by reducing the quality of execution or practice of an event due to lack of warmup.

* *Attitudes towards automation* could best be expressed as an indication of frustration over the operation of automated systems in a non user friendly environment, although improvements in the human-machine interface would probably reduce this feeling to some extent. Wherever introduced, automation has not been uncritically accepted by those who are meant to operate it. Some aspects of automation are accepted while others are rejected (in some cases because operators did not operate the equipment acceptably in the real world environment). Acceptance of automation may also be affected by factors related to the culture of the organization to which employees belong. Poor relationships with management, employee perceptions of having had no choice in the decision to accept automation, and lack of involvement in the development of automation are other examples of factors that may negatively affect the acceptance of automation. These factors may operate independently of the quality of the automation provided to the employees.

* *Motivation and job satisfaction* involve problem areas such as loss of the controller's feeling of importance, the perceived loss in the value of professional skills, and the absence of feedback about personal performance. Many operators feel that their main source of satisfaction in their job lies in its intrinsic interest to them. They believe that the challenge of the job is one of the main reasons they enjoy their profession. A takeover by automation to the point that job satisfaction is reduced can lead to boredom and general discontent.

* *Over-reliance on automation* occurs because it is easy to become accustomed to the new automated systems' usefulness and quality. A tendency to use automation to cope with rapidly changing circumstances may develop even when there is not enough time to enter new data into the computer. When things go wrong, there may also be a reluctance by the human to discard the automation and take over.

* *Systematic decision errors*: Humans may depart from optimal decision-making practices, particularly under time pressure or other stress. The existence of human biases may further limit the

ability of humans to make optimal decisions. One approach to reduce or eliminate biased decision-making tendencies is to use automated decision-making aids at the time decisions are required. In such a system, humans adopt one of two strategies: accept or reject the machine recommendation. Although benefits of automated decision making aids are theoretically evident, they still remain to be conclusively demonstrated.

* *Boredom and automation complacency* may occur if a major portion of air traffic management is completely automated, and human operators are lulled into inattention. In the particular case of complacency, humans are likely to become so confident that the automatic systems will work effectively that they become less vigilant or excessively tolerant of errors in the system's performance.

* *Automation intimidation* results in part because of an increase in system components. The result is a reliability problem, since the more components there are, the more likely it will be that one will fail. However, humans remain reluctant to interfere with automated processes, in spite of some evidence of malfunction. This is partly due to inadequate training and partly to other pressures.

* *Distrust* normally occurs because the assessment of a particular situation by the human differs from the automated system. If the system does not perform in the same manner as a human would do, or in the manner the controller expects, it can lead to either inappropriate action or concern on the part of the human. This can also occur if the human is not adequately trained. Distrust can be aggravated by flaws in system design which lead to nuisance warnings.

* *Mode confusion and mode misapplication* are results of the many possibilities offered by automation, as well as of inadequate training. It is possible with a new computer technology for the controller to assume that the system is operating under a certain management mode when in fact it is not.

* *Workload* The advance of automation has been based partly on the assumption that workload would be reduced, but there is evidence to suspect that this goal has yet to be achieved. In the air traffic control environment, additional working practices such as data entry/retrieval methods may actually increase workload. For example, merely automating certain aspects of an ATC system will not necessarily enable the air traffic control officer to handle more traffic. Automation should be directed at removing nonessential tasks, thereby allowing the controller to concentrate on more important tasks, such as monitoring or directly controlling the system.

* *Team function* The team roles and functions in automated systems differ from those which can be exercised in manual systems. As an example, controllers in more automated systems are more self-sufficient and autonomous and fulfil more tasks by interacting with the machine rather than with colleagues or with pilots. There is less speech and more keying. This affects the feasibility and development of traditional team functions such as supervision, assistance, assessment and on-the-job training. When jobs are done by members of closely coordinated team, a consensus about the relative merits of individual performance can form the basis not only of professional respect and trust but also of promotions or assignments of further responsibilities.

Technology-centred approach in the automation of highly advanced technologies such as the nuclear power plant industry, chemical industry, civil aviation, space technology, etc., resulted in accidents with a great loss of lives and property. Basically, such accidents were an outcome of human-machine incompatibilities. Since the technology was easily available, engineering-based solutions to human error were implemented without due consideration of human capabilities and limitations. Technology-centred automation may be based on the designer's view that the human operator is unreliable and inefficient, and so should be eliminated from the system. However, two ironies of this approach have been identified: one is that designer errors can be a major source of operating problems; the other is that the designer who tries to eliminate the operator still leaves the operator to do the tasks which the designer does not know how to automate. To this we can add the fact that automation is not, after all, infallible and usually fails in mysterious and unpredictable ways. It is for this reason that there are increasing calls for a human-centred approach which takes all the elements, and especially the human element, into due consideration. Hard lessons have been learned in the automation of aviation systems in the past. Cockpit automation stands as an example. However, in cockpit automation, we can now say that — albeit with notorious exceptions — there is a return to human-centred automation, which is a positive and encouraging trend strongly endorsed by ICAO. It is hoped that lessons learned in the past are applied to all new advanced technology systems so that same mistakes will not be committed.

Oceanic flight data processing system with conflict prediction and resolution. (Picture provided courtesy of CAE Electronics Ltd.)

9. A view of the future system from the air

Philippe Domogala

Introduction

To illustrate what the future CNS/ATM system might be like for the pilots, we will examine a typical, imaginary flight on a Boeing 747 of a major airline from Los Angeles to Frankfurt in the year 2010. This route is used because it covers most kinds of airspace one would find around the world. We begin at a high density airport, then enter dense continental airspace above the United States, moving on to remote areas of Northern Canada, then to Oceanic airspace above the North Atlantic Ocean, entering dense continental airspace again above the United Kingdom and Germany, descending through a dense terminal airspace, finally landing in Frankfurt, Germany in poor visibility conditions.

Strategic planning by the aircraft operator

The Flight Plan as defined by the airline (i.e., flight number, type of aircraft, route requested, date and time of departure, destination, alternate aerodrome and equipment (avionics) available on the aircraft) is transmitted to air traffic services (ATS) or ATM as it is now known, up to 6 months in advance, although the flight plan will only be specifically dealt with about 48 hours prior to the time of operation. During the prior months, the automated ATM system continuously searches its data bases in order to identify areas of potential problems such as excessive air traffic demand for a given portion of airspace or airport for the day of operation.

This particular flight plan, which we shall call GLOBAL AIR 123, arrives in the computer of the Central ATFM facility of the Continental United States in Washington D.C. The flight plan is then automatically transmitted to the Canadian and European ATFM centralized units, and each of them assesses the impact of that particular flight on their systems.

As additional flight plans from other airlines, military and general aviation are entered into the system, a possible overload is detected above London for a particular period the following morning, the time our aircraft is expected to be entering that airspace. Automatically, a message is sent to our GLOBAL Airline asking for a possible shift in the requested departure time so that GLOBAL 123 will arrive in the London airspace either before or after the overload period. As our Airline is a scheduled carrier, publishing its schedule months in advance and GLOBAL 123 is a scheduled flight, unlike chartered or business flights which can be more flexible, GLOBAL cannot accept and expressed its preference to take a reroute on the day of operation.

About 48 hours prior to departure, the flight plan is automatically brought up and specifically treated by the ATM system in order to prepare the system for the day of operations. By this time, the airspace management plan (AMP), describing the military activity on the day of the flight (i.e., opening and closure times of the danger and restricted airspaces), is activated and distributed.

On a daily basis, the ATM system automatically prepares a pre-tactical plan for the following two day period, thereby distributing equally delays and reroutings, in order to avoid the congested areas detected in the Continental United States. GLOBAL Airlines is automatically sent a pre-notification message for their flight 123, indicating what is to be expected in two days.

When re routing is inevitable, extra fuel can be added and other factors can be planned for by flight operations.

Three hours prior to departure

About three hours prior to departure, a final departure slot, which consists of a departure window of about ten to fifteen minutes, valid for a particular routing, is sent to the airline. When the Captain of GLOBAL 123 arrives in the briefing room, about one hour prior departure, he is assured of a time and a route for which his flight has been accepted into the system.

He immediately sees that his company has accepted a reroute and that adds about 200 miles on to the preferred route. He is also given a complete and computerized flight plan with the estimated time over all significant points along the route. This data has already been pre-programmed into the FMS aboard the aircraft via data link from GLOBAL's flight operations. Additionally, all of the notices to airmen (NOTAMs) in force for the entire flight have been added to the memory of the FMS.

When our pilot enters the cockpit, about 30 min. prior to departure, all preliminary checks have already been accomplished by remote computers, again via data link, and the final status check of aircraft systems has been printed out on a small printer located between the two pilot seats. This will be one of the few pieces of paper in the cockpit, and this mainly for backup reasons. Most of the paperwork such as approach plates, navigation charts, NOTAMs, weather reports, load sheets, etc., that were to be found in cockpits of the past, are now placed into ground data bases, all available to the aircraft computer via air to ground data link.

The ramp manager enters the cockpit about ten minutes before door closing time in order to input the latest load figures such as the final number of passengers and weight and balance information. He presses a button and all of the information is also automatically linked to the airline's operation's central computer.

Start up

The doors are finally closed at 1640 in the afternoon as GLOBAL 123 is planned to depart at 1705. The latest slot time received on the data link monitor in the cockpit is 1720. if the aircraft is not ready by that time, it will be issued a fifteen minute delay. Also appearing on the monitor is the average taxi time needed to reach the runway in use, precisely at the departure time issued by ATC. The pilot now knows he will need about 22 minutes to reach the active runway and plans his startup and "push back" accordingly.

Our pilot decides to ask for start up thirty minutes prior to the departure time to avoid any potential problems. To do this he simply enters the time of 1650 in the data link entry device and presses the entry button, thereby automatically indicating to the TWR controller that he will be ready to start-up at 1650. About five seconds later, a small symbol appears on the data link display indicating to the pilot that the TWR controller has received his request.

About two minutes after this exchange of information between the TWR and the pilot, without intervention by either, the pre-departure clearance appears on the data link display in the cockpit. It also indicates to the pilots, that the requested startup time has been approved, the taxiway to follow to reach the active departure runway, and where to hold for further clearance. The message ends with the words that all pilots are pleased to see and hear: *no delay expected.*

Our two pilots adjust the updated parameters received by their FMS. Remember there is no longer any need to mention SSR transponder codes to squawk as part of the clearance. The Mode S Transponder they are equipped with allows each individual aircraft to retain its own permanent code, known by all the ATM systems around the world.

Taxiing

The operators and airline personnel on the ramp had also received a copy of the pre-departure clearance in their trucks and were preparing for push-back.

Ready to push-back captain? blared over the loudspeaker in the cockpit. Our pilots realized this was the first voice message received since they seated themselves in the cockpit. The captain was still responsible for taxiing, however, even this was greatly facilitated by GNSS in 2010. A map of the airfield was displayed on one of the liquid crystal displays (LCDs), with the cleared taxi route to be followed outlined as a thick blue line. Our captain, who is now 54 years old, reminisces of the Boeing 747-100 series, which required a good deal of hands-on activity for flying and taxiing. He knows that, if he wanted, he could switch on the ground computer and the aircraft would automatically taxi to the required position, but as it was a beautiful, sunny day, he decides to get a feeling for the aircraft and do the taxiing himself.

A few minutes later the monitor flashes: *Able takeoff 1714?* Both pilots look at each other and realize that another aircraft had probably missed his slot. Fortunately, GLOBAL 123 had started up a little early for just such a possibility. After a quick calculation the pilot determines that he can make the departure slot. The copilot presses the *affirmative* button on the data link entry device and almost immediately a revised message is received from the TWR clearing GLOBAL 123 via a new route along with the revised time of take off.

When our aircraft arrives at the threshold of the departure runway, the pilot presses the *ready for takeoff* selection on the data link entry device and almost immediately receives the reply on the display: *monitor 119.7.* The frequency 119.7 was then selected on the VHF voice radio telephone box, and voices began to fill the cockpit. All critical instructions for the next few minutes will be via voice. This is still the case for all critical phases of flights such as for takeoff and landing clearances, changes of heading and altitude changes necessary for separation purposes. Voice was still preferred to data link because it was much faster.

Also, for human factors, or human-machine interface reasons relating to such things as intonation of voice, degree of urgency and party line effect, voice was considered to be an added safety aspect in busy terminal areas for now. Research is still being done; however, in the meantime, this has become the standard practice.

Takeoff

A few minutes after arriving at the threshold of the active departure runway, the data link display flashes to indicate that the departure clearance is ready: *after take off climb to 5,000 feet on the standard instrument departure (SID) 5A, Monitor 118.75*

A few moments later the TWR controller verbally instructs the flight: *GLOBAL 123 cleared to line up runway 05 and hold. Roger, lining up GLOBAL 123* replies our pilot.

Then the awaited call comes from the TWR : *GLOBAL 123 cleared for take off runway 05*. The pilot looks down at the display to read the latest wind information, automatically collected and transmitted via data link from the various anemometers located around the airfield. Information on possible windshear is also available and is coupled with an automatic digitized audio warning.

As our aircraft accelerates down the runway, the pilots scan the cockpit displays checking the parameters: *V1, ...Vr* states the copilot and the Boeing 747 gently lifts off. Finally airborne, immediately all the computers on board the aircraft register the new actual time of departure as 1716 and revise all navigation parameters accordingly. The actual time of departure is also transmitted automatically to the airline operations and air traffic control, updating all ground systems.

Climb-out to cruising altitude

The pilot engages the autopilot when passing 2,000 ft. and climbing. The copilot selects the frequency of the departure controller: 118.72. Immediately, the enroute clearance appears on the screen clearing the flight directly to a point on the Canadian border thousands of miles away, based on the Free Flight concept, with the clearance to cruise initially at flight level 320 (e.g., thirty two thousand ft.). All aircraft flying in the airspace above 15,000 ft. over the continental United Sates are now certified as having the ability to maintain a specific navigation accuracy based on the required navigation performance concept. The same is true for communications. This accuracy has allowed the Free Flight concept to be implemented and almost all flight operations at these upper levels are via direct routings. When aircraft come within proximity of each other, ground computers, in coordination with flight computers, dynamically adjust the path of the aircraft.

GLOBAL 123s flight plan indicates that FL340 was initially requested, and 360 after two hours. The FMS will remind the pilots and ATC of this when the time comes. As our pilots scan the displays and keep an eye out the window, all appears normal, when suddenly a loud voice shouts in the cockpit: *TRAFFIC, TRAFFIC*. On the airborne collision avoidance system (ACAS) display, which is now much larger and easier to read than the earlier systems of the 1990s, both pilots see light, general aviation aircraft flying above and below their Boeing 747. But it turned out to be a false alert, and soon the symbol returned to *normal* traffic. Although it appeared to be a very busy situation, the pilots were not worried. This was normal and the ACAS provided enough warning for them to avoid the traffic.

They were however, focusing on another blip that the flight management computer pointed out on the situational awareness display which they felt could pose a problem for them by possibly coming too close. This display was a separate function from the ACAS. Civil aircraft were now equipped with automatic dependent surveillance-broadcast (ADS-B) transmit and receive capabilities. This enhanced situational awareness allowed the pilot and flight management computers to have a more active role in air traffic management. This was also a technical development that led to an ATM operational concept which eventually allowed the Free Flight and other ATM advanced concepts to be implemented, leading to greater autonomy of flight.

Passing 18,000 ft, the data link display flashes with the message: *Monitor centre 135.32*. They tune in the frequency assigned and the data link display immediately confirms their initial clearance: *Continue climb to FL320, direct to PONTY* (PONTY is a newly designated point on the United States/Canadian border). The pilots relax and a few minutes later the voice over the loudspeaker fills the cockpit again: *GLOBAL 123 for separation turn right heading 035*. The captain immediately replies verbally: *Turning right heading 035, GLOBAL 123*. Again, when quick action has to be taken, even though its the result of a computer alert, he uses voice communications instead of data link.

After a short while the controller asks the pilot of our flight to resume his own navigation, meaning that he can proceed back to the cleared route as the conflicting traffic has passed 5.6 miles of thc left side of the aircraft.

Cruising

A few more routine messages concerning frequencies to monitor are exchanged via data link and our aircraft is now cruising at 32,000 ft. About an hour and one half later the FMS alerts the pilots that the aircraft, now lighter as fuel is burned off, would fly more economically at FL360. At the same time, the pilot is notified that the FMS will notify ground computers that the aircraft would like to climb. The captain presses a few keys on the data link entry device and transmits: *Requesting FL360*. After a few seconds the reply from ATC comes back and is presented on the display: *Unable next 30 Min. Be advised light to moderate turbulence reported above FL350 area north of ROCKY*. The controller notes the information being sent to the aircraft and makes no intervention. Turbulence reporting had been tremendously enhanced since the introduction of automatic downlinking to ATC of wind and temperature parameters by all aircraft, at regular intervals. They decide to wait for their climb.

Until now, all the navigation had been done using GNSS augmented with the United States' wide area augmentation system (WAAS). The FMS is also updated by the few ground navigational aids (navaids) that still exist, however, the GNSS accuracy was much greater. The navaids remained more as an eventual backup to GPS and GLONASS signal failures.

Navigation, once the main source of worry for pilots, had now become one of the easiest of the pilots tasks.

Our flight is now crossing the Canadian border and in a few minutes will enter the remote area of the Northern Territories where ATM will be provided using non-radar, computerized techniques at first, and will then be based on ADS. A warning in the cockpit advises the crew that they must remain alert as their position will no longer be detected or followed by radar, but by the automatic transmission of their calculated position by the on-board FMS. Again, all aircraft flying in this airspace had to be equipped for ADS. An error in their FMS and the downlinked position would be wrong. In such a situation, the ATM system would not immediately be able to detect the deviation, and the safety of theirs and other flights could be put at risk.

The pilot remembers that only a few years ago at the current position of his aircraft, the cockpit would be filled with static noises and loud voices as all aircraft had to transmit their position on high frequency radios every 10 minutes. Now with data link, all was silent, leading to a greater degree of safety and more accuracy. Another big advantage is that with ADS, separation standards were lowered to 30 miles in areas where they would have been as much as 20 minutes, which resulted in accommodation of more aircraft on the most direct, great circle route, also closer to their optimum altitudes.

The pilots decide to remind ATM that they would like to climb to FL360. Another request is sent via data link and a clearance is received: *Cleared to climb to FL 360,* again, this clearance required no controller intervention.

A few hours later GLOBAL 123 is turned over to oceanic control prior to entering oceanic airspace over the North Atlantic. Once again the data link display delivers the oceanic crossing clearance. The crossing track was optimally determined that morning based on the winds for the day and the aircraft has been cleared on a track very close to the optimum track requested. The pilots were pleased with the route that had been chosen.

Still under ADS based control, the pilots knew that they were entering an area where many aircraft were concentrated, 30 miles apart and restricted to mach number speeds, all wanting to be on the most optimum routes based on the wind factor.

It should be a relatively peaceful flight now as they enter oceanic airspace and they settle in for the crossing. The pilot continues to monitor the speed of the aircraft as there is another aircraft 30 miles behind at the same level and speed, even though the FMSs of the two aircraft are now in communication with each other and will automatically monitor and adjust the speeds of the two aircraft to ensure the separation is maintained. They know that in about three hours, when they enter the interface area with London control, they will become very busy. They also know however, that they will be back under full radar service by then, and five miles separation will be in effect.

The congestion was getting worse in London in 2010. Everybody wanted to fly to Heathrow, but the Environmentalist Party had become so powerful recently, that no new runways nor airports could be built in the area. Furthermore, a stringent nighttime curfew was introduced which concentrated even more flights in less hours.

Significant efforts had been made in the last fifteen years to improve airspace capacity, however, in some countries, airport capacity and airport access was becoming worse.

Arrival

GLOBAL 123 was rerouted numerous times via data link communications, to avoid some congested areas above the United Kingdom. The flight was then transferred to the Eurocontrol Upper Area Control Centre (UACC). This was one of six UACCs in Europe which covered the whole of the European airspace above 29,000 ft. These multinational centres were operating on an RNAV systems concept where the whole airspace was considered as being available, thereby allowing aircraft to fly direct routes, freeing them from a fixed route system.

As our flight enters the Eurocontrol UACC airspace, it is cleared direct on an RNAV track to a fixed point 200 miles further along the route of flight. Later, it was alerted to offset its track 5 miles to the right. This is one technique used to accommodate more aircraft on their preferred routes, by offsetting the aircraft 5 miles from each other on parallel tracks. As in the United States, much of this tactical decision-making was accomplished by ground computers. Deviation monitoring systems were in place, both on the ground and on board aircraft, as reaction times in case of deviation were extremely short when aircraft were cleared to cruise this closely to each other. Although this technique and the associated technology was rather old (DC10s had this capability in the early 70s) it could only be made widely available a few years ago, when the three remaining aircraft manufacturers (Boeing, Airbus and Tupolev-Douglas) finally agreed to standardise the logic of their FMS's and auto pilots. Additionally, the pilots and FMS could monitor all aircraft in their vicinity on their ADS-B situational display, allowing easy identification of nearby traffic that might blunder.

The concept of required navigation performance-5 (RNP-5) was now fully in place in Europe and, together with the 1,000 ft. separation at all altitudes, maximum use of the airspace could be achieved. Aircraft unable to comply with RNP 5 or with modern altimetry enabling 1000 ft. vertical separation above 29,000 ft, were relegated to the lower airspace to fly along fixed routes as before.

Descent and approach

At about 200 miles from destination, the data link monitor displays the first clearance in preparation for landing: *Reduce speed to Mach .76*. This is the first sequential restriction they will receive from the Frankfurt approach metering system.

The system was developed to prevent holding of aircraft in the air, and to sequence aircraft coming from different directions at different speeds and altitudes, for landing on a particular runway. It did not increase capacity as such, but allowed a smoother, safer and less stressful operation regulation for the approach controllers.

The next restriction received was based on politics. About 50 miles prior to the German border, our flight was issued a clearance to descend to 27,000 ft. This was about 60 miles prior to what the FMS calculated would be the most economical descent for the aircraft. As in the United Kingdom, the Environmental Party in Germany was in power in many "Landers" and had passed a law restricting the number of flights above 30,000 ft. in the belief that this would help to preserve the ozone layer and reduce global warming. The first consequence was that all flights inbound for German airports were instructed to cross the German border below 30,000 ft. to allow more overflights, as there were restrictions on the number of flights allowed to operate in German airspace above 30,000 ft.

The decent was programmed by the pilots on their FMS and a message was received to *monitor 131.68*. A strange frequency, but since the introduction of 8.33 KHz channel spacing was introduced because of frequency congestion problems in Western Europe, the pilots were getting used to it, especially in the upper airspace.

The data link display flashed again: *Icing possibility between 14,000 and 6,000 ft. over Frankfurt area.* This information was being passed automatically to all aircraft approaching Frankfurt. In fact, the information was collected automatically from a satellite distribution system of world weather information, and then processed for data link transmission to aircraft. The copilot pressed the two deicing buttons on his overhead panel and another message came in: *Cologne airport closed for next twenty five minutes due snow removal.* Our pilots take notice as Cologne was their first alternate. The Captain switches the FMS to choose the second alternate which is Dusseldorf. This automation eases considerably the workload of the pilots.

A controller's voice fills the cockpit: *GLOBAL 123, turn left Heading 190 to intercept MLS runway 25L, visibility down to 200m.* The data link display confirms what the controller stated verbally: *RVR 200m in snow showers.* Years ago this would have meant a diversion, but today all landing systems were certified to Category III B conditions meaning that visibility could be zero and aircraft would be able to land as long as the pilots were trained and aircraft were equipped with fully automated landing systems, which required no human intervention.

The MLS proved to be very good for this particular kind of weather. While GNSS approaches were the norm for most of the rest of the world, and ILS was slowly becoming obsolete, Europe presented a unique situation in that weather conditions were often very poor and, in the 1990s, a decision was made to rely more heavily on MLS as it was believed that a problem would exist concerning Cat IIIB approaches using GNSS only. This was still an issue in 2010 even as the technology improved.

Landing

The pilot made his first request for landing clearance via voice communications: *Frankfurt Approach, this is GLOBAL 123, we can make a curved track direct to 25 left if it helps.*

Affirm, replied the controller also using voice communications. *You are number two behind an Airbus 340 on short final coming from the north, 6 miles out presently.* The copilot saw the traffic "electronically" on the situational awareness display and both pilots spotted the Airbus. *Traffic on display, 8 miles on our one o'clock, 700 feet below.* The controller immediately replies: *This is correct GLOBAL 123, maintain at least 2.5 miles behind the Airbus 340, continue approach.* The copilot manually adjusts the autopilot to reduce speed and then selects a specific curved trajectory in order to automatically maintain a distance of 2.5 miles behind the Airbus.

Everything proceeded smoothly, and without further input, the controller calls again: *GLOBAL 123 cleared to land on runway 25 left.* Our aircraft lands via a fully automated, curved MLS approach, coupled with the FMS on-board the aircraft, using GNSS as a backup in this case. After touchdown the auto pilot is disconnected and the pilots tune in the ground system. Automatically, the map of Frankfurt airport comes on the display with the route to follow to their assigned gate depicted as a bright blue line. As the visibility was very poor and the pilots could hardly see the blue taxi way lights on either side of the aircraft, they selected the auto-taxi feature, and the GNSS system guides and taxis the aircraft automatically to the gate.

That is the end of our flight. The pilots relax as the computers on board the aircraft make some final calculations and data link this information to the airline operations centre 7,000 miles away.

Training issues

Much of the new technology that GLOBAL 123 made use of would have been introduced sometime between 2000 and 2010, requiring many controversial changes to the way that pilots and controllers work and interact, not only with each other, but also with the machines they operate.

Training, testing, involvement, simulation and retraining was the key to winning acceptance of all these new systems and procedures by pilots and controllers.

The introduction of data link introduced the concept of the silent cockpit which was a revolutionary change for both controllers and pilots at the turn of the century. For the pilot, monitoring the performance of automated systems became one of his most important tasks.

Change was resisted at first, however as efforts were made by the system engineers and the civil aviation authorities toward recognition of the benefits, and those benefits were fully realized and accepted by those who would operate within the new CNS/ATM systems, fears were quickly alleviated and the operators of the system became the greatest supporters of the new systems and contributed largely to its success.

Human factors issues

Pilots, like controllers, have to be very careful and alert when using the new method of communication (i.e., data link). As data link communication became more prevalent, it was soon realized that over the years, voice communications had developed many important qualities that had a direct impact on safety; qualities that could never be reproduced, such as voice intonation, the use of non-standard phraseology, which often indicates indecision, incomprehension, confusion or even fear. Now one had to rely solely on the printed text. Most importantly, however, was the loss of the "party line" effect which allowed everyone tuned into a particular frequency to hear all of the transmissions on that frequency. This had the effect of keeping everyone aware and to some degree, involved. Many incident and accidents in the past were most likely avoided because someone on the frequency "got it right" and pointed out the mistake.

Overall, however, data link offered an exactness, enhanced accuracy and speed along with a reduction in confusion, which compensated for any deficiencies.

An additional, significant aspect for pilots, was that they had a tendency to become over-reliant on the automation, and gradually lose their basic flying skills. More and more flights were now being operated automatically. Navigating had become so easy, and GNSS had become so accurate that pilots were slowly losing their navigational knowledge, formally, the most basic and important of skills. The reality, however, was that these skills might be badly needed in the event of system and then backup system failure. To keep those basic skills in an almost totally automated environment was a serious problem in 2010. Simulation was compensating to some degree for this problem.

The final point to made regarding human factors issues deals with ADS. Many pilots had the tendency to believe that ADS was the same as radar, since the service provided by ATC was almost the same. Pilots have to be constantly reminded that ADS is a dependent surveillance system, totally reliant on the accuracy of the information input and transmitted by the aircraft systems to the ground. Unlike radar, however, an error could not yet be detected by ground systems in a timely manner and corrected. In fact, more monitoring was required when using ADS than when using any other kind of surveillance, and humans, as it turned out, are very bad monitors, preferring decision-making. Pilots were no exception to this general rule, and training programmes now focus intensely on this issue.

Benefits

* The main benefits for the pilots operating in the fully integrated, global CNS/ATM system was increased safety;

* Vast amounts of airspace, previously controlled via procedural means using voice reports, was now positively controlled using ADS;

* More accurate navigation using GNSS offered constant reassurance to pilots;

* Increased availability of backup communications systems assured pilots that there would be almost no chance of losing communications completely;

* Access to numerous data bases via data link (e.g., weather, systems diagnosis, maps, charts, etc.) relieved pilots of many meticulous duties and increased safety;

* Dynamic reroutings allowed more direct routes at optimum flight profiles, reducing flight time and delays;

* Communication errors were reduced because of data link;

* ATM units responded more quickly to pilots requests; and,

* Pilots have access to more airfields for diversions, using GNSS landing capabilities.

10. A view of the future system from the ground

Philippe Domogala

Introduction

To illustrate what the work of the air traffic controller will be like in a well developed future CNS/ATM system, we will view the air traffic control (ATC) system from the point of view of the air traffic controllers who operate the system. First, from the perspective of a tower controller at a busy international airport, then from that of an approach controller and also a controller at a busy national enroute control centre, or area control centre (ACC) to use the correct name. We will view the system through the eyes of an oceanic controller, then from those of a newly created automatic dependent surveillance (ADS) controller covering a remote area, and finally through the eyes of a controller at a multinational upper area control centre. Imagine all of this taking place in the year 2010.

A look behind the scenes

Although air traffic controllers, (or managers) are in the limelight, one must realize that a great deal of work to keep the CNS/ATM system functioning is performed by other people, behind the scenes, through each of the stages of a controlled flight.

Flight plan processing (FPP)

The first in line are the people and systems that receive and process the flight plans. In the past (don't forget, we are now in the year 2010), airline pilots and pilots of smaller airplanes, or general aviation pilots, physically filed their flight plans, usually in triplicate, in aeronautical information service (AIS) offices. Those pieces of paper, or flight plans, then had to be retyped in the form of telexes or on teletypewriter machines, so that they could be sent to all ATS facilities which would be affected by the flight.

Today, in the year 2010, a handful of large centralized flight plan processing (FPP) units, all linked together via direct lines, using the aeronautical telecommunications network (ATN) protocol, accomplish the task of filing, checking and transmitting the flight plan information. In fact, the airlines send their schedules in the form of stored flight plans automatically to these FPP units months in advance of actual flight times. Private pilots and smaller airlines that operate on less rigid schedules are also linked to these central systems by way of dedicated data bank terminals, located at most airports around the world. Those States with highly developed telecommunications systems allow flight plans to be filed via commercially available telephonic interfaces, such as the MINITEL in France, or even through the Internet.

This efficient and automatic flight plan filing system, combined with the introduction of Mode S radar and transponder equipment, with its 16.7 million codes, allowing a dedicated individual transponder code for each and every registered aircraft in the world, allows instant and accurate flight plan processing from just about anywhere on Earth.

The automated processing and distribution of flight plans is a tremendous milestone for ATM because now, each and every ATM unit affected by a flight, anywhere in the world, is assured of automatically receiving accurate details of that flight, well in advance of actual flight times. This information allows not only proper planning of staff, but a significantly reduced workload for controllers and an enhancement of safety as ATM units are operated in a more efficient and effective manner.

Air traffic flow management (ATFM)

The next element of the CNS/ATM system operating behind the scenes that we will look at is that of air traffic flow management (ATFM). Here again, in all of the world in the year 2010, there are only a few centralized large

ATFM units (e.g., Brussels, Washington, Moscow, Cairo, New Delhi and Sydney) that interact with each other and predict airspace and airport capacity overloads, also integrating weather information in order to plan for eventual reroutes or to determine the best track system based on the winds.

By 2010, the ATFM system is finely tuned to ensure that air traffic controllers are never confronted with situations where air traffic demand is greater than capacity, or where there are more aircraft than can safely be handled. Furthermore, the controller will never have to hold aircraft because of poor weather conditions, or determine the quickest and most economic routes for pilots in the unlikely event that diversions are necessary.

Airspace management (ASM)

The next system operating behind the scenes that we will investigate is that of airspace management, and the associated ASM unit, where all of the requests for reservations of airspace are collected and processed days before an actual operation is to take place. These requests come mainly from the military, but also from other users, such as aircraft manufacturers, flying clubs or air show organizers.

With all airspace reservation and other use requests now centralized and coordinated, it is possible to design a four dimensional map (i.e. including the times of activity) of all the airspace in a given region during a particular day. This allows flights to be cleared flights directly to their destinations as soon as this is possible and to use more flexible altitudes above and below reserved airspace during the periods in which these airspaces are actually being used.

This new concept of ASM allows a much more efficient use of the airspace at any given time and increases the overall efficiency and effectiveness of the ATM system.

Finally

Of course the system engineers and technicians had developed a data network over the years, operating between ATM facilities using the ATN protocol. Radar data is now shared between States, via area networks, voice communications are relayed via satellite when necessary, and all flight plan and other information such as weather and notices to airmen (NOTAMs) is shared and redistributed on request among any ATM facility. This technology allows a degree of redundancy for ATM facilities that never existed before. For example, if a system fails in one ATM unit, data could be obtained almost immediately from a neighbouring unit, minimizing the disturbance.

This cooperation between neighbouring States has been unprecedented. In fact, the only problem encountered in the last years had been the final acceptance by all the ATM authorities in the world, of a single programming language for ATM computational systems, and a single protocol of data exchange. It had been a long and vicious battle, where each large, industrialized State defended its own interests, but finally common sense prevailed and the second generation ATN (i.e., ATN2) was approved and introduced. The early benefits of the common protocol were such that within a few years most States had adopted or retrofitted their systems to work with the ATN2 protocol.

Beginning the shift with the tower (TWR) controller

Our TWR controller entered the TWR building about twenty minutes before he was due to relieve his colleague in the "hot seat". This was a common practice for controllers. These twenty minutes allowed the relieving controller to read the notices to airmen (NOTAMs) in force for that day, to become acquainted with any changes in the ATM system, to prepare for eventual outages, to study the weather forecast and the list of planned departures and arrivals for the airport for that day. It is also common to "spring" the controller on shift to an early out. Controllers are generally punctual employees, however, they all intensely disliked being relieved late after a long difficult shift and, in a close knit community such as that of the controller, you generally got back what you gave.

Arriving at the top of the TWR, the controller first looks at the planning schedules for the next hour on the supervisors desk, exchanges a few hellos and jokes with his colleagues, puts on his top of the line, glare proof Ray-Bans, and proceeds without further delay to stand behind and observe the colleague he is scheduled to relieve.

He scans radar screens associated with his sector or area of responsibility, reads a few lines of data link information and studies the approach radar display, all in order to mentally determine exactly where the traffic is and to build a mental picture of the traffic situation.

Since our controller is now able to scroll down the data link display and read all the previous messages exchanged between the pilots and the controller he is relieving, unlike before, there is no need for lengthy verbal briefings before taking over the sector and sitting in the hot seat. Unlike a car, or many other ground based systems, you cannot just stop the flights in midair every time you change controllers, or some other contingency arises. The change over procedure resembles two jugglers in a circus act, exchanging batons while they are in the air, while one juggler replaces the other from below. Despite all other technological advances, this was still the change over procedure in 2010.

Now our controller is in the hot seat at the take off position. He follows the progress of his aircraft on the monitors and displays, and also visually monitors the situation outside through the huge TWR windows. He follows the preparatory work done by his assistants, the Ramp controller and the Ground controller. He accomplishes this monitoring function simply by looking at the data link displays, following the exchanges of messages between the various controllers, pilots and assistants.

Most pre-departure clearances and enroute clearances are automated. This automated system also allocates the frequencies for pilots to monitor and the routes to taxi on to reach the threshold of the departure runway, thus avoiding repetitive inputs for each flight. The latest weather is also automatically collected from various sources including satellites, and broadcast via data link. And of course, the sequencing of departures is automated and aircraft begin lining up as the busy period approaches.

Since flight plan processing is now centralized and aircraft are individually coded based on their discrete Mode S addresses, transfers and coordination between TWR and approach (APP) control are also largely automated. In fact, very few verbal exchanges take place, except of course in special circumstances like emergencies and special requests.

Our TWR controller looks outside and uses voice communications to clear aircraft to line up and take off, separating them from landing aircraft and the numerous crossing vehicles and other vehicular traffic at the airport. After numerous efforts at automating the critical decision making processes of the controller, by 2010, a human-centred approach to automation was finally implemented by most States, with the fully automated system put off for another day. Our TWR controller, therefore, continues to make critical decisions. After an initial scepticism, most controllers accepted and enjoyed the newly automated systems, as they assisted them in carrying out their routine tasks, while also helping them with the less critical decision making tasks and keeping a vigilant lookout for conflicts. Based on past experiences, most manufacturers and ATM providers and authorities have ensured that the air traffic managers are fully involved in the system design, development and implementation. This has ensured installation of equipment and systems that are easily assimilated by the managers and more readily accepted.

These improvements have made the controllers job much less stressful, while at the same time requiring that they bear the responsibility for managing a large complex ATM system, which is why the name air traffic manager had slowly evolved, replacing the more technical one of air traffic controller. The training for the new managers is significantly different than in the past. The new hires are now required to learn how to manage and interact with large complex systems and to understand such concepts as intelligent processing and human factors.

The latest upgrade introduced in our TWR had also been largely accepted and praised by the managers as they were responsible for the concept in the first place, arising from an air traffic managers suggestion. It was a simple system based on the global navigation satellite system (GNSS), using Mode S transponders, to track every aircraft, car and truck and, in fact anything that could move on the airfield. The system then automatically warns, via an audio and visual signal, the air traffic managers when someone or something is not where it is supposed to be, for instance, crossing an active runway without a clearance or taking a wrong taxiway. This system is also used to give instructions to aircraft during periods of low visibility, thereby eliminating the need for expensive ground movement radar. The pilots, in fact, have access to the same information on their situation display.

The approach controller

Here also in the approach control facility, the controllers are pleased with the new systems. This particular unit makes use of the latest Monopulse Mode S Radar as its main surveillance tool, however, over the past few years, automatic dependent surveillance (ADS) was also introduced and integrated into the system. For the first time, controllers could see aircraft behind the mountain range, inside the valley, as well as at very low altitudes. This was virtually impossible using the older line-of-sight radar equipment.

At the "departure" position, controllers read the exact clearances issued to aircraft by the TWR on the data link display, and can therefore detect problems much earlier than before. Depending on the standard instrument departure (SID) route in force and also on the ever increasing noise abatement procedures applicable at that time of the day, the controller adjusts the rate of release or, minimum departure interval between flights, to ensure a smooth transition into the enroute system.

When the controller modifies the departure intervals, this information is automatically transmitted to the ATFM Unit, which in turn adjusts departure slots for all aircraft waiting to depart from this particular airfield.

The main difference for the approach controller in 2010 from past times however, is experienced while working at the "arrival" position . With the introduction of sophisticated, intelligent metering devices, a new form of arrival sequencing has been put in place. Aircraft now adjust their speed as early as 200 miles from the airport in order to sequence themselves, based on information automatically transmitted from the metering function of the ATM system, to the arriving aircraft via data link. Aircraft arrive at the entry fix into the approach area with roughly between two and four miles spacing from each other, irrespective of the direction they are coming from.

Being human-centred, the automated system allows the controller to decide on modifications to the minimum distance between flights for arrival. In extremely good conditions he could reduce this distance to as little as 2.5miles.

During the approach phase, most of the routine transmissions were via data link, however, radar vectoring and altitude changes during the final phases of the flight are accomplished via voice communication. The approach controller, like his TWR counterpart, enjoys this one-on-one dialogue with the pilots during this critical phase of the flight. He also enjoys the security afforded to him by the short-term conflict alert system that informs him if two aircraft are about to become too close to each other. This helps him with his concentration and reduces the stress levels of the job, knowing that the system was there to assist him if he inadvertently missed out on something vital.

The main problem still encountered by approach controllers in 2010 was the non standardization of landing aids. After strenuous battles in the 1990s between industrialized States on a single, standardized future landing aid, there was in 2010, three systems remaining in use: The GNSS based system, used in the United States, the States of the former Soviet Union, most of Africa and Central and South America and parts of Asia, the microwave landing system (MLS) used in Western Europe and a few States in Eastern Europe. The standard instrument landing system (ILS) also continues to be used everywhere else and especially in States where frequency protection issues could be easily resolved.

The airlines had long ago given up hopes of being able to equip with one type of receiver and have since equipped their fleets with all three types of receivers. All of the receivers feed a single display in the cockpit allowing a standard procedure regardless of the type of approach used. The certification of GNSS category IIIB continues to be pursued, and many European States plagued by persistent low visibility have therefore equipped themselves with MLS.

For controllers, approach paths could be either straight in when ILS is used, curved when MLS is in operation, or any style when GNSS approaches are in operation.

The enroute controller (continental airspace)

Unlike the approach and TWR controllers, the continental enroute controller's job has changed less dramatically with the introduction of automation. He now typically controls air traffic over greater areas with the help of automation, in altitude blocks between fourteen thousand and thirty thousand feet using Mode S radar and ADS for surveillance and also ADS as a back up. Fewer ground-based navigational aids are used now as most aircraft navigate using the GNSS. There is still a mix of propeller and jet traffic as general aviation has proliferated over the past twenty years and many aircraft still fly on fixed ATS routes, although overall there is a much greater degree of freedom to proceed on direct routings than in the past.

Required navigation performance 5 (RNP 5) is the norm in most heavily flown continental airspaces in 2010; therefore, in these areas, any route could be flown by an aircraft if it is certified to RNP 5. This means that there is a certain amount of assurance that the aircraft will remain on its assigned track. The RNP concept led to reductions in separation between aircraft, and was possible only because of the newer technologies. However, the heavy mix of different types of aircraft, military requirements, and the non-availability of sophisticated on-board equipment and the expense of equipping to RNP 5 certification standards by much of the general aviation community have meant that, in reality, the amount of flexibility previously hoped for concerning direct routings was somewhat limited.

The enroute controller now has very effective new tools, like short term conflict alert (STCA), medium term conflict alert (MTCA) and deviation monitoring software. Additionally, the availability of trajectory prediction, based on the continuous downlinking of aircraft parameters, allows much better preplanning than before.

Aircraft could also downlink such information as their top of descent and requests for altitude changes and diversions, thereby allowing an interaction and even negotiation between on board and ground systems.

The introduction of one thousand ft. vertical separation above 29000 ft. has also meant that some older types of aircraft were now excluded from flying in the upper airspace of many continental airspaces. This meant that many military aircraft as well as those older Russian and Chinese built aircraft were not allowed in this reduced vertical separation minima (RVSM) airspace because of the demanding navigation and height keeping accuracies required. The excluded aircraft were now forced to operate in the middle altitude band, together with high performance turbo props, thereby causing new and unforeseen difficulties for our enroute controller.

The introduction of metering devices for some approaches also unexpectedly complicated the work of the enroute controller. With more and more airfields equipping themselves with such devices, the enroute controller was faced with numerous speed adjustments taking place in his airspace, sometimes confusing and in opposition to the controllers planning, due to the fact that these adjustments were coming from different systems, serving different runways at different airports. These problems are being worked on and a fix is expected in the near future, based on adjustments to the software in the United States' system and resectorization techniques in European airspace.

Unfortunately, in 2010, the situation for the enroute controller, while vastly different than in past times, continues to be difficult. This has raised interesting industrial, social issues amongst the controllers and the authorities.

The upper area control centre (ACC) controller

In contrast to the continental ACC controller, the upper ACC controller works over very large areas. In some regions, his area of responsibility may cover four or five countries and typically all of the airspace above 30,000ft. He could now use the whole airspace without much restriction and without having to base his control responsibilities on fixed, ground-based navigational aids, or States' FIR boundaries and other sectorization problems as the airspace is overall, more efficiently organized, based on the regional concept of ATM.

Using Mode S Radar, with ADS as a back up, the upper ACC controller now clears aircraft along any desired track, or he can create at will a set of parallel routes, using the offset possibilities afforded by the advanced flight management systems on-board aircraft operating in the upper airspace, thereby allowing aircraft to occupy the same, optimum cruising levels if so requested.

The need for radar vectoring is greatly reduced, as is the controllers workload. In most cases, four dimensional contracts are negotiated between FMSs and ground computer systems (i.e., a track, an altitude, and a time to be at a given point, which may be hundreds of miles further along), while controllers pass on to the pilot the responsibility to execute that contract. These contracts are usually conflict free, however, separation assurance is reinforced with STCA and MTCA and track deviation monitoring systems, alerting both the pilot and the controller via data link and audio signal, when an aircraft deviates from the plan. Conflict resolution systems are also being introduced whereby the ground computer system offers resolution possibilities to the controller when a conflict is detected.

Unfortunately, the airspace above some areas of the United States, Europe, Japan and other parts of Asia has now become so saturated that the use of certain techniques such as free flight in the United States and direct flight in Europe, could not be fully implemented on a 24 hour basis.

The introduction of 1000 ft vertical separation above 29,000 ft. at first created a tremendous increase in capacity in the upper system, however, this problem was partially negated with the merger of the three main aircraft engine manufacturers in 2003. Now most new types of aircraft were equipped with the same engines, and therefore all requesting the same, most economical cruising altitudes. As a result, most traffic was now concentrated in a small, high altitude band, and the much expected gains of RVSM were in fact reduced.

The introduction of 8.33 KHz channel spacing between VHF frequencies was initially carried out in the upper airspace over Western Europe and, as a result, most ATC facilities were able to open new sectors fairly rapidly, as enough frequencies were available to do so, a sharp contrast with the 90's where sometimes an air traffic control unit had to wait up to two years to obtain a frequency.

Radio failures were now almost non existent. Each aircraft now has at least two or three VHF radios, 2 data link systems, 1 satellite communications system (SATCOM), and two Mode S transponders, also capable of transmitting and receiving data link messages.

Flight plan data also flows rapidly between ATC facilities, and with the introduction of ground data link question marking ability making further use of intelligent processing (e.g., can I do this ? or, what if I do this ?), most telephonic conversations between controllers were now reduced to a bare minimum.

The oceanic and remote area controller

Previously, the oceanic or remote area controller worked with less adequate equipment, or in less developed regions. Often, the less developed States had no funding for advanced equipment. With the introduction of ADS, this situation has changed significantly.

This controller had to adapt, perhaps more rapidly than most others to new technology. From an environment where he obtained voice position reports and used pieces of paper to keep track of aircraft, where poor quality, high frequency radios were used for communication, and then only with radio operators, and separation between aircraft was ten and twenty minutes, his world changed to a high tech one, where colour situation displays were introduced for monitoring and controlling aircraft, data link with SATCOM voice as back up were implemented for communications and twenty mile longitudinal and 1000 ft. vertical separation minima was implemented.

This revolution did create a few problems. Airlines and State aviation administrations pushed hard to have all procedural airspace in the world declared as ADS airspace as soon as possible, in order to reduce separation standards and introduce more direct routes. However, it turned out that making full use of ADS often required a high level of automation, including powerful computers to process the flight plan data, ground to ground datalink capabilities between ATM units, fast processors to provide the tools necessary for safety assurance, communications facilities, ADS displays and expensive retraining of staff. This often meant higher than expected investment costs, not always possible for many States. A cost-recovery plan, based on route charges was therefore put into place, based on the forecast amount of overflying traffic. The plan clearly showed that some ACCs could not adequately update their facilities and expect a return on their investment sufficient to cover investments costs. In some places overflying charges were drastically raised to cover investment costs. In many cases, this was clearly worth it for the airlines, for instance, over Siberia or China. The amount of flight time saved was sufficient that operators were willing to make the necessary investments to automate and upgrade the Russian Federation ATM facilities. There were other examples, however, where airlines were faced with exorbitant route charges if they used an airspace, often forcing a decision to avoid the airspace completely, even if it meant additional mileage. The problem was a long standing one and of a serious enough nature that ICAO was now planning a Divisional type worldwide meeting to deal with the it.

When ADS was in place in remote areas and over the oceanic regions, however, and operated effectively, it allowed controllers working within the new ATM system to plan flights more effectively and to expedite them with the same efficiency as would normally be available in dense continental airspace, using radar. Response time to pilot requests in such areas was usually around forty to sixty seconds, as opposed to ten or twenty minutes, as was previously the case. Optimal routings were created through areas previously inaccessible (e.g. Siberia, Sahara, Amazon, China) and dynamic rerouting was now available over heavily traversed oceanic regions such as in the South Pacific. As more and more aircraft were now able to fly according to the most efficient and economical route, it was often the case that aircraft previously spread over large areas, were now concentrated over smaller areas based on the most economical route between major city pairs.

The introduction of SATCOM also revolutionized the way that pilots and controllers now communicate above these remote and oceanic areas. Unfortunately, full availability of SATCOM also was fairly expensive for the ATM authority, meaning that some States decided to retain use of high frequency radio in 2010. High frequency voice radio was also still used over the Polar Regions where SATCOM was still not available.

Management of emergencies

With the introduction of CNS/ATM also came a revolution in the way that search and rescue (SAR) services were provided. Although satellites for the provision of search and rescue for detecting aircraft and locating aircraft in distress had been around for quite a few years, it was only recently that GNSS broadcasting was introduced in the emergency locators aboard all aircraft, thereby allowing the search and rescue authorities to pinpoint with extreme accuracy, the exact location of any of these locator beacons when activated.

Additionally, pilots experiencing emergency situations could simply press a button on their Mode S transponder and automatically broadcast not only their emergency status, but also all of their aircraft parameters, including number of passengers on board, the nature of the emergency, fuel on board, etc. This information provided valuable information to the rescue services. It also left the voice frequencies free for vital communications with the pilot. Furthermore, using ADS capabilities, navigation to any suitable diversion runway could be coordinated in seconds, as well as landing on any of those runways regardless of weather conditions, due to the fact that GNSS approaches could be conducted almost anywhere in the world.

Training

As is the case with pilots, training issues for air traffic managers are critical with the introduction of the new technologies. This fact was often lost during the initial stages of CNS/ATM implementation. It was soon realized that unlike airplanes, one cannot stop the ATM system, remove the old equipment, send the controllers to school, install the new equipment, develop and publish new procedures, and when the stage is set a few months later, inaugurate the new system and reopen the facility with freshly trained controllers.

Furthermore, it is neither socially nor financially feasible to build a completely new system alongside the old one, recruit and train new controllers for this and on the changeover day, discard the old material and people. The realization was eventually arrived at by most authorities that the progression to the new CNS/ATM systems would be evolutionary and the same people operating the older systems would have to be constantly retrained to assimilate new procedures and systems while newer equipment would have to be introduced in an evolutionary manner, as aircraft continue flying.

The notion of training is important, but affects geographical regions differently. In areas where older procedural control techniques were still used, training for systems such as ADS and in ATM procedures continues to take place. The reality in 2010 is that ADS procedures are more complex than radar procedures, mainly because as ADS systems and procedures are installed and used for air traffic control in the same way as radar, there is often no similar backup system in case of failures and complex contingency procedures had to be established for such circumstances.

To transfer from the most basic, procedural type air traffic control, to air traffic control based on ADS, requires a great deal of training, motivation and development of new skills on the part of the controllers or managers. For most, however, the challenge was accepted worldwide after some serious negotiation, coordination and compromise with the organizations and associations representing the controller's interests. This usually involved agreements leading to a much more active involvement in training, planning and implementation of systems and procedures by the controllers. Eventually, the status of the profession on the whole was significantly raised along with the wages of course, as the job evolved into a highly professional one.

Human factors issues

The introduction of all this new technology created some serious human factors issues and some of these are covered in previous chapters in this text. The most important issue was based on the result of years of system engineering that led, initially, to an overload of information for the controllers, for the simple reason that it became available. The controllers for their part applied a great deal of pressure aimed at a reduction in the amount of information displayed to them as well as a reduction in the number of possibilities offered to them by the system.

Additionally, with the tremendous increase in the amount of traffic in the recent years in many areas, the alerting devices (i.e., STCA, MTCA and track deviation devices) all devised to assist the controllers, soon were considered hyperactive by the controllers and were often disengaged, defeating the initial alerting purposes. This was often the case in the past with the introduction of early models of new technologies (i.e., ground proximity warning systems (GPWS) and airborne collision and avoidance systems (ACAS)).

Another significant human factors issue had been the tendency by the younger generation of controllers, and also of pilots in some cases, to over rely on the automated alerting devices, thereby reducing the level of vigilance that should always be exercised in a human centred automated system. Enhanced training techniques, often computer-based, appeared to be alleviating this shortcoming.

Additionally, reducing the separation minima both laterally and vertically had resulted in a grouping of aircraft that had never been experienced before. This meant that the time left to counteract, or react to a contingency was extremely small. A reliance on airborne collision avoidance systems (ACAS) to solve these problem was still the subject of controversy in 2010, mainly as the result of several reports of near midair collisions, sometimes emotionally, by the media.

Finally, the role of the controller in general had changed from being proactive (i.e. detecting problems and solving them) into a more reactive one (i.e. following a list of problems presented by the system and choosing options in given situations, as proposed by the system). This, combined with the increasing use of data links and an impersonal communications medium, created some psychological difficulties for the older generation of controllers as they saw their professions evolve before their eyes and new, younger people take on the computerized tasks with little difficulty while they stood hopelessly by, longing for bygone days.

Benefits

To sum up the benefits from the perspective of the operators or managers of the newer CNS/ATM systems, the following list has been compiled by some operators of the system:

* With the introduction of ADS, safety had been greatly improved in regions which were previously uncontrolled or were controlled using procedural techniques;

* Flight times were in some cases significantly reduced with the opening of new routes above remote areas such as Siberia, the Sahara, China, etc.;

* The introduction of RNAV combined with RNP5, enhanced FMS and data link capabilities and ATM automation, drastically reduced the workload of controllers, often eliminating the need for radar vectoring;

* Datalink had reduced voice communications to a minimum, especially regarding the need for transmission of routine information;

* ADS allowed aircraft to be detected in areas previously unreachable by radar as well as providing an accurate back up to radar in case of failure;

* The main benefit of the new system, from the perspective of the controller, however, was the worldwide activation of ground/ground and air/ground datalink using a common protocol. This allowed controllers to be aware of almost all traffic coming to them, and with extreme precision. The downlinking of aircraft parameters also allowed the development of a global, integrated ATM system. It was evolving toward a truly functional and interoperable system where flight operation, ATS, ASM and ATFM were being integrated into a total system, all aimed at reducing the workload of the controllers, and therefore increasing safety, capacity and job satisfaction.

PART D

REFERENCES

References

Abudaowd, H. (1993), *Working Paper presented at the Fourth Meeting of the Special Committee for the Monitoring and Co-ordination of Development and Transition Planning for the Future Air Navigation System (FANS(II)4-WP/77),* International Civil Aviation Organization, Montreal.

Aeronautical Satellite News, (1992), "United proposes route revolution", Aeronautical Satellite News, no. 31, December 1992-January 1993.

Aeronautical Satellite News, (1993a), "French ADS trial detailed", *Aeronautical Satellite News*, no. 34, June-July 1993.

Aeronautical Satellite News, (1993b), "Italy set for satellite ADS trial", *Aeronautical Satellite News*, no. 35, August-September 1993.

Aeronautical Satellite News, (1994a), "Inmarsat to decide on allocation of navigation transponders", *Aeronautical Satellite News*, no. 41, October-November.

Aeronautical Satellite News, (1994b), "New trials herald round-the-world ADS", *Aeronautical Satellite News*, no. 38, April-May.

Aeronautical Satellite News, (1995), "Inmarsat allocates navigation transponders", *Aeronautical Satellite News*, no. 43, February-March.

Airport Forum, (1995), "ICAO satellite navigation panel sets goals", *Airport Forum,* 1/1995.

Air Transport International, (1995), "Ground-to-air-control", *Air Transport International,* 1 - 7 March 1995.

Aleshire, W. (1994a), "FANS 1 flight management system", *Journal of Air Traffic Control*, vol. 36, no. 4, October-December.

Aleshire, W. (1994b), "FANS 1: Fast track to the future", *Aeronautical Satellite News,* no. 39, June-July.

Andresen, E.S. (1991), "Automation key to future ATM system", *ICAO Journal*, vol. 46, no. 12, December.

Andresen, E.S. (1993), "Air traffic management supported by future CNS components can cope with all traffic scenarios", *ICAO Journal*, vol. 48, no. 10, December.

Arkind, K.D. and Medis, P. (1996), "Operational concept demonstration of FANS 1 ATC in India", *ATC Systems,* vol. 2, no. 1, January/February.

Berger, P. (1984), *Summary of international cooperation in aviation up to 1945*, International Civil Aviation Organization, Paris.

Billings, C.E. (1989), "Toward a human-centred automation philosophy", *Proceedings of the Fifth International Symposium on Aviation Psychology*, Columbus, Ohio.

Campbell, F.A., and Salewicz, G.W. (1995), "Satellite-based system takes navigation into the 21st century", In M. Blacklock (Ed.), *International Civil Aviation Organization: 50 years global celebration,* International Systems and Communications Ltd., London.

Canada, Mexico and the United States of America, (1995), *Canada, Mexico, United States of America CNS/ATM implementation and transition plan*, Federal Aviation Administration, Washington.

Castro-Rodriguez, F. (1990), "Automatic dependent surveillance promises increased efficiency and safety", *ICAO Journal,* vol. 45, no. 9, September.

Castro-Rodriguez, F. (1994), "ICAO standards for ADS and digital data links are keys to implementation of ADS-based ATC systems", *ICAO Journal,* vol. 49, no. 9, November.

Challinor, J.G. (1994), "Australia's ADS and data link programme", *Working Paper presented at the Third Meeting of the Automatic Dependent Surveillance Panel* (ADSP/3-WP/10), International Civil Aviation Organization, Montreal.

CNS Outlook, (1995), "DGPS appears feasible for CAT III landings, FAA says", *CNS Outlook,* vol. 3, no. 12, September.

Cole, H.W. (1990), "Mode S, data link, ACAS, a status report", *The Controller*, vol. 29, no. 12, February.

Daly, K. (1994a), "Making the connection", *Flight International*, 25-31 May.

Daly, K. (1994b), "Regionals offered FANS capabilities", *Flight International*, 20-26 July.

Daly, K. (1996), "Talk this way", *Flight International*, 7-13 February.

Del Balzo, J. (1993), "Satellite-based systems will soon replace ILS, MLS in the United States, delegates told", *Airport Forum,* 4/1993.

Della Rocco. P, (1990), "Selection of air traffic controllers for automated systems: Applications from today's research", In J.A. Wise, V.D. Hopkin, & M.V. Smith (Eds.), *Automation and Systems Issues in Air Traffic Control,* NATO ASI Series F, vol. 73, Berlin, Springer-Verlag.

Department of State, United States of America, (1944), *Proceedings of the International Civil Aviation Conference*, United States Government Printing Office, Washington.

Department of State, United States of America, (1948), *Aspects of United States participation in international civil aviation*, United States Government Printing Office, Washington.

Diez, D. (1993a), "ADS/SSR data integration study", *Working paper presented at the Fourth meeting of the Special Committee for the Monitoring and Cooperation of Development and Transition Planning for the Future Air Navigation System (FANS Phase II) (FANS(II)/4-WP/42)*, International Civil Aviation Organization, Montreal.

Diez, D. (1993b), "Improved surveillance is possible by combining data from different systems", *ICAO Journal*, vol. 48, no. 2, March.

Dorfler, J. (1994), "Realization of early operational benefits based on the use of satellite navigation", *Working Paper presented at the First Meeting of the Global Navigation Satellite System Panel (GNSSP/1-WP/7),* International Civil Aviation Organization, Montreal.

Eng, C.H. (1993), "Singapore's advanced ATC system is scheduled to enter operation in 1994", *ICAO Journal*, vol. 48, no. 2, March.

Eurocontrol, (1995a), "Development of the European air traffic management system (EATMS)", *Information paper presented at the Thirty-Seventh Meeting of the European Air Navigation Planning Group (EANPG/37-IP/2),* International Civil Aviation Organization, Paris.

Eurocontrol, (1995b), *EATMS Mission, Objectives and Strategy Document (MOSD)* (Eurocontrol Doc No. 93 70 41), Eurocontrol, Brussels, Belgium.

Evans, D.S. (1994), "Industry could speed ICAO rule-making", *Aviation Week and Space Technology*, vol. 141, no. 18, October.

Featherstone, D.H. (1993), "Specialist providers offer cost-effective approach to obtaining ATS satellite communications", *ICAO Journal*, vol. 48, no. 5, June.

Federal Aviation Administration, (1994a), "Use of global positioning system (GPS) in the U.S. national airspace system (NAS)", *Information Paper presented at the Special European Regional Air Navigation Meeting (SP EUR-IP/8),* International Civil Aviation Organization, Montreal.

Federal Aviation Administration, (1994b), "Status of U.S. R&D with the global positioning system, *Information Paper presented at the Special European Regional Air Navigation Meeting (SP EUR-IP/7),* International Civil Aviation Organization, Montreal.

Federal Aviation Administration, (1994c), "Civil aviation use of the U.S. Global Positioning System", *Information Paper presented at the Special European Regional Air Navigation Meeting (SP EUR-IP/9),* International Civil Aviation Organization, Montreal.

Federal Aviation Administration, (1995), "Free flight", *Information bulletin for the 41st Paris international air show,* Federal Aviation Administration, Washington, D.C.

Flight International, (1995), "Germany embarks on GPS testing", *Flight International*, 18-24 January.

Freer, D.W. (1986a), "An aborted takeoff for internationalism - 1903 to 1919", *ICAO Bulletin Part 1*, March.

Freer, D.W. (1986b), "A convention is signed and ICAN is born - 1919 to 1926", *ICAO Bulletin*, May.

Freer, D.W. (1994), "ICAO at fifty years: Riding the flywheel of technology", *ICAO Journal*, vol. 49, no. 7, September.

Galotti, V.P., Kornecki, A. (1991), "Knowledge engineering for an air traffic expert system", *36th Annual Air Traffic Control Association Proceedings,* Arlington, VA.

Galotti, V.P. (1992), "An expert air traffic control teaching machine: Critical learning issues", *In J.A. Wise, V.D. Hopkin & P. Stager (Eds.), Verification and Validation of Complex and Integrated Human-Machine Systems*, Berlin, Springer-Verlag.

Galotti, V.P, Heijl, M. (1996), "ICAO working to establish an integrated, global air traffic management system", *ICAO Journal*, vol. 51, no. 4, May.

Gribben, W.J. (1991), "Commercial satellite communications undergo Pacific trials", *ICAO Journal,* vol. 46, no. 3, March.

Gupta, R. (1994), "New funds crucial to ICAO tech support", *Aviation Week and Space Technology*, vol. 141, no. 18, October.

Hartman, R. (1992), "Flight tests of GPS and GLONASS equipment", *The Journal of Air Traffic Control,* vol. 34, no. 1, January-March.

Heijl, M. (1994), "CNS/ATM road map for the future portrays an evolutionary process", *ICAO Journal,* vol. 49 no. 4, May.

Hopkin, V.D. (1989), "Man-machine interface problems in designing air traffic control systems", *IEEE Proceedings: Air Traffic Control,* vol. 77, no. 11, November.

Howell, C. (1994), "The Australian GNSS transition plan", *Working Paper presented at the First Meeting of the Global navigation Satellite System Panel (GNSSP/1-WP/52),* International Civil Aviation Organization, Montreal.

Hughes, D. (1995), "Free flight sparks international debate", *Aviation Week and Space Technology,* July.

Hurn, J. (1989), *GPS: a guide to the next utility*, Trimble Navigation, Sunnyvale, California.

International Air Transport Association, (1991), "Airlines foresee FANS utilization in oceanic airspace by end of decade", *ICAO Journal*, vol. 46, no. 12, December.

International Air Transport Association, (1995a), "The Boeing FANS 1 System", *Working paper presented at the Meeting on Transit ATS Routes through the airspace of the Eastern part of the ICAO European Region, including Middle Asia (TARTAR/2 WP/20),* International Civil Aviation Organization, Paris.

International Air Transport Association, (1995b), "CNS/ATM implementation", *Working paper presented at the Meeting on ATS routes through the airspace of the Eastern part of the ICAO European Region, including Middle Asia (TARTAR/2 WP/19),* International Civil Aviation Organization, Paris.

International Civil Aviation Organization, (1947), *What is ICAO*, International Civil Aviation Organization, Montreal.

International Civil Aviation Organization, (1951), *Memorandum on ICAO*, International Civil Aviation Organization, Montreal.

International Civil Aviation Organization, (1953), *Memorandum on ICAO*, International Civil Aviation Organization, Montreal.

International Civil Aviation Organization, (1958), *Memorandum on ICAO*, International Civil Aviation Organization, Montreal.

International Civil Aviation Organization, (1984), *The Air Traffic Services Planning Manual* (1st ed.) (Doc. 9426), International Civil Aviation Organization, Montreal.

International Civil Aviation Organization, (1985a), *Report of the Second Meeting of the Special Committee on Future Air Navigation Systems (Doc.9458, FANS/2),* International Civil Aviation Organization, Montreal.

International Civil Aviation Organization, (1985b), *International standards and recommended practices, Annex 10 - Aeronautical Telecommunications (4th ed. of Vol. II),* International Civil Aviation Organization, Montreal.

International Civil Aviation Organization, (1988a), "FANS Committee proposes a consolidated global CNS plan", *ICAO Bulletin*, June.

International Civil Aviation Organization, (1988b), *Report of the Fourth Meeting of the Special Committee on Future Air Navigation Systems (Doc.9524, FANS/4),* International Civil Aviation Organization, Montreal.

International Civil Aviation Organization, (1988c), *Secondary surveillance radar - Mode S data link (Circular 212-AN/129),* International Civil Aviation Organization, Montreal.

International Civil Aviation Organization, (1989), "Fundamental human factors concepts", *Human Factors Digest No. 1 (Circular 216-AN/131),* International Civil Aviation Organization, Montreal.

International Civil Aviation Organization, (1990), *Automatic dependent surveillance, (Circular 226-AN/135),* International Civil Aviation Organization, Montreal.

International Civil Aviation Organization, (1991a), *Report of the Tenth Air Navigation Conference (Doc. 9583, AN-CONF/10),* International Civil Aviation Organization, Montreal.

International Civil Aviation Organization, (1991b), *Manual of the aeronautical telecommunication network (1st ed.) (Doc. 9578, AN/935),* International Civil Aviation Organization, Montreal.

International Civil Aviation Organization, (1991c), *Manual of area navigation (RNAV) (1st ed.) (Doc. 9573-AN/933),* International Civil Aviation Organization, Montreal.

International Civil Aviation Organization, (1992a), *Facts about ICAO,* International Civil Aviation Organization, Montreal, Canada.

International Civil Aviation Organization, (1992b), *Strategy for the implementation of area navigation (RNAV) (4th ed.) (EUR Doc. 001, RNAV/4),* International Civil Aviation Organization, Paris.

International Civil Aviation Organization, (1993a), *Report of the Fourth Meeting of the Special Committee for the Monitoring and Coordination of Development and Transition Planning for the Future Air Navigation System (Doc. 9623, FANS(II)/4),* International Civil Aviation Organization, Montreal.

International Civil Aviation Organization, (1993b), *Regional Office Manual, Third Edition*, International Civil Aviation Organization, Montreal.

International Civil Aviation Organization, (1994a), *Catalogue of ICAO publications and audio visual aids, 1994 edition,* International Civil Aviation Organization, Montreal.

International Civil Aviation Organization, (1994b), *Memorandum on ICAO: The story of the International Civil Aviation Organization,* International Civil Aviation Organization, Montreal.

International Civil Aviation Organization, (1994c), *Information kit for ICAO's 50th anniversary,* International Civil Aviation Organization, Montreal.

International Civil Aviation Organization, (1994d), *Report of the Fifteenth Meeting of the All Weather Operations Panel (AWOP)*, International Civil Aviation Organization, Montreal.

International Civil Aviation Organization, (1994e), "Global navigation satellite system (GNSS): Review of the report of the GNSSP/1 meeting", *Working paper of the Air Navigation Commission (AN-WP/6965),* International Civil Aviation Organization, Montreal.

International Civil Aviation Organization, (1994f), *Summary of discussions of the Third Meeting of the North Atlantic Automatic Dependent Surveillance Development Group (NAT ADSDG/3),* International Civil Aviation Organization, Paris, France.

International Civil Aviation Organization, (1994g), "ADS related activities by Norway and Sweden", *Working Paper presented at the Third Meeting of the Automatic Dependent Surveillance Panel (ADSP/3-WP/49),* International Civil Aviation Organization, Montreal.

International Civil Aviation Organization, (1994h), "Transition to GNSS in Canada", *Working Paper presented at the First Meeting of the Global Navigation Satellite System Panel (GNSSP/1-WP/13),* International Civil Aviation Organization, Montreal.

International Civil Aviation Organization, (1994i), "Human factors in CNS/ATM systems", *Human Factors Digest No. 11 (Circular 249-AN/149),* International Civil Aviation Organization, Montreal.

International Civil Aviation Organization, (1994j) *Manual on Required Navigation Performance (Doc. 9613-AN/937),* International Civil Aviation Organization, Montreal.

International Civil Aviation Organization, (1995a), " IATA's user driven plan for CNS/ATM implementation in the Eastern part of the ICAO European Region", *Working Paper presented at the Second Meeting on ATS transit routes through the airspace of the Eastern part of the ICAO European Region, including Middle Asia (TARTAR/2-WP/12),* International Civil Aviation Organization, Paris.

International Civil Aviation Organization, (1995b), "Annual review of developments on technical and operational aspects of the ICAO CNS/ATM systems", *Working paper of the Air Navigation Commission (AN-WP/6986),* International Civil Aviation Organization, Montreal.

International Civil Aviation Organization, (1995c), "Annual review of developments on technical and operational aspects of the ICAO CNS/ATM systems", *Working paper of the ICAO Council (C-WP/10176),* International Civil Aviation Organization, Montreal.

International Civil Aviation Organization, (1995d), "General review of activities on CNS/ATM systems implementation in the Russian Federation", *Information paper presented at the Thirty-Seventh Meeting of the European Air Navigation Planning Group (EANPG/37-IP/9),* International Civil Aviation Organization, Paris.

International Civil Aviation Organization, (1995e), " The Inmarsat-3 navigation transponders", *Information paper presented at the Thirty-Seventh Meeting of the European Air Navigation Planning Group (EANPG/37-IP/10),* International Civil Aviation Organization, Paris.

International Civil Aviation Organization, (1995f), "Draft report of committee A to the meeting on agenda item 3", *Working paper presented at the ICAO Special Communications/Operations Divisional Meeting (SP COM/OPS/95-WP141),* International Civil Aviation Organization, Montreal.

International Civil Aviation Organization, (1995g), "ICAO special communications/operations meeting crafts 20-year strategy for all-weather non-visual approach and landing operations", *ICAO news release (PIO 3/95),* International Civil Aviation Organization, Montreal.

International Civil Aviation Organization, (1995h), "ICAO CNS/ATM Implementation Committee (CAI)", *Working paper presented at the Thirty-Seventh Meeting of the European Air Navigation Planning Group (EANPG/37-WP/31),* International Civil Aviation Organization, Paris.

International Civil Aviation Organization, (1995i), "Overview of ATM-related tasks and proposed executive summaries", *Working paper of the Air Navigation Commission (AN-WP/7024),* International Civil Aviation Organization, Montreal.

International Civil Aviation Organization, (1995j), "Required total system performance", *Technical work programme of the Air Navigation Commission, (ANC Task No. RAC-9501),* International Civil Aviation Organization, Montreal.

International Civil Aviation Organization, (1995k), "ATM requirements for communication", *Technical work programme of the Air Navigation Commission, (ANC Task No. RAC-9502),* International Civil Aviation Organization, Montreal.

International Civil Aviation Organization, (1995l), "ATM requirements for navigation", *Technical work programme of the Air Navigation Commission, (ANC Task No. RAC-9503),* International Civil Aviation Organization, Montreal.

International Civil Aviation Organization, (1995m), "ATM requirements for surveillance", *Technical work programme of the Air Navigation Commission, (ANC Task No. RAC-9504),* International Civil Aviation Organization, Montreal.

International Civil Aviation Organization, (1995n), "Airspace infrastructure planning", *Technical work programme of the Air Navigation Commission, (ANC Task No. RAC-9505),* International Civil Aviation Organization, Montreal.

International Civil Aviation Organization, (1995o), "Inter-operability and functional integration of flight operations, ATS, ATFM and tactical ASM", *Technical work programme of the Air Navigation Commission, (ANC Task No. RAC-9510),* International Civil Aviation Organization, Montreal.

International Civil Aviation Organization, (1995p), "ATFM systems", *Technical work programme of the Air Navigation Commission, (ANC Task No. RAC-9507),* International Civil Aviation Organization, Montreal.

International Civil Aviation Organization, (1995q), "ATFM procedures", *Technical work programme of the Air Navigation Commission, (ANC Task No. RAC-9508),* International Civil Aviation Organization, Montreal.

International Civil Aviation Organization, (1995r), "Integration of regional ATFM systems and procedures", *Technical work programme of the Air Navigation Commission, (ANC Task No. RAC-9509),* International Civil Aviation Organization, Montreal.

International Civil Aviation Organization, (1996), "Annual review of developments on technical and operational aspects of the ICAO CNS/ATM systems", *Working paper of the ICAO Council (C-WP/10368),* International Civil Aviation Organization, Montreal.

International Coordinating Council of Aerospace Industries Association, (1991), "Aerospace industry ready to help implement global solution", *ICAO Journal*, vol. 46, no. 12, December.

Jackman, F. (1994), "Bilateral issue spurs ongoing debate", *Aviation Week and Space Technology*, vol. 141, no. 18, October.

Klass, P.J. (1993), "KGLS technique sharpens GPS accuracy", *Aviation Week and Space Technology*, May 10.

Kotaite, A. (1992), "Council President reports to the 29th Assembly on an eventful three years", *ICAO Journal*, vol. 47, no. 11, November.

Kotaite, A. (1993), "Mechanism to provide implementation coordination and assistance under active consideration", *ICAO Journal*, vol. 48, no. 10, December.

Kotaite, A. (1994), "Message from the President of the ICAO Council", *First Choice Canada*, vol. 12 no. 2, Summer.

Kotaite, A. (1995), *Address by the President of the ICAO Council to the Special Communications/Operations Divisional Meeting, (SPCOM/OPS/95),* International Civil Aviation Organization, Montreal.

Kuranov, V. (1994), "GLONASS system current status and prospects of its development and use as the GNSS system component", *Working Paper presented at the First Meeting of the Global navigation Satellite System Panel (GNSSP/1-WP/42),* International Civil Aviation Organization, Montreal.

Learmount, D. (1993), "Satellite signposts for African skies", *Aeronautical Satellite News,* no. 34, June 1993.

MacLean, D. (1994), "Canadian ADS trials in the Gander airspace", *Working Paper presented at the Third Meeting of the Automatic Dependent Surveillance Panel (ADSP/3-WP/48),* International Civil Aviation Organization, Montreal.

Malescot, D., and Chenevier, E. (1993), "Application of a methodology for determining separation standards in an ADS-ATC system", *Working Paper presented at the Eighth Meeting of the Review of the General Concept of Separation Panel (RGCSP/8-WP/26)*, International Civil Aviation Organization, Montreal.

McInnis, R. (1995), "Going forward", *IFALPA International Quarterly Review,* September.

Morrocco, J. D. (1996), "Europe tests core of new ATM systems", *Aviation Week and Space Technology*, June 3.

Nickum, J.D., (1992), "Air carrier utilizes Satcom for ATC communications", *ICAO Journal*, vol. 47, no. 1, January.

Nordwall, B.D. (1993), "Norway's broadcasters to carry D-GPS signal", *Aviation Week and Space Technology*, August 2.

Nordwall, B.D. (1996), "Asia/Pacific leads FANS 1 progress", *Aviation Week and Space Technology*, August 12.

Norris, G. (1994a), "Watching the clock", *Flight International*, November-December.

Norris, G. (1994b), "FAA allows primary GPS over oceans", *Flight International*, December 1994-January 1995.

O'Keefe, B. (1991), "Flight path to the future", *ICAO Journal*, vol. 46, no. 12, December.

O'Keefe, B. (1993), "System development enters new phase with detailed regional planning underway", *ICAO Journal*, vol. 48, no. 10, December.

Olsen, D. (1993), "Global NavCom: A new ATM era", *Airports International,* September.

Ott, J. (1994), "Open skies haunts Chicago Convention", *Aviation Week and Space Technology*, vol. 141, no. 18, October.

Peterson, D. (1995), "Status of WAAS procurement", *Presentation made to the Second Meeting of the Eurocontrol Satellite Navigation Applications Sub-Group (SNA/CON/MIN.03),* Eurocontrol Experimental Centre, Bretigny-Sur-Orge, France.

Pozesky, M.T. (1993), "National airspace system precision approach and landing system (NASPALS) plan", *Working Paper presented at the Fourth Meeting of the Special Committee for the Monitoring and Coordination of Development and Transition Planning for the Future Air Navigation System (FANS(II)/4-WP/32),* International Civil Aviation Organization, Montreal.

Price, F. (1994), "Development of FANS 1 implementation plan", *Working Paper presented at the Third Meeting of the Automatic Dependent Surveillance Panel (ADSP/3-WP/56),* International Civil Aviation Organization, Montreal.

Price, H.E. (1990), "Conceptual system design and the human role", *MANPRINT*, In H. R. Booher (Ed.), Van Nostrand Reinhold, New York.

Rochat, P. (1995), *ICAO State letter: Notification of IATA, GLOBAL NAVCOM 95 symposium and exhibition promoting global implementation of the ICAO CNS/ATM system (M 4/1-95/9),* International Civil Aviation Organization, Montreal.

RTCA, (1995), *Report of the board of directors' select committee on free flight*, RTCA Incorporated, Washington, D.C.

RTCA, (1996), *Interim draft of the minimum aviation system performance standards for automatic dependent surveillance broadcast (ADS-B)*, ARINC Incorporated, Annapolis, Maryland.

Ruitenberg, B. (1994), "ICAO CASITAF 2", *The Controller,* vol. 33, no.4, December.

Russian Department of Air Transport, the Russian Commission for Air Traffic Regulation and the U.S. Federal Aviation Administration, (1994), "Russian Civil aviation system safety evaluation", *Joint report of the Russian Department of Air Transport, the Russian Commission for Air Traffic Regulation and the U.S. Federal Aviation Administration,* The Federal Aviation Administration, Washington, D.C.

Ryals, L.Z. (1993), "Aeronautical telecommunication network (ATN) validation and implementation initiatives", *The Journal of Air Traffic Control,* June, 1993.

Ryan, F. (1994), *Working Paper presented at the First Meeting of the Global Navigation Satellite System Panel (GNSSP/1-WP/21),* International Civil Aviation Organization, Montreal.

Signargout, L. (1995), "Aircraft builder emphasizes flexible approach to implementation of satellite-based systems", *ICAO Journal*, vol. 50, no. 2, March.

Sweetman, B. (1994), "Airlines get connected", *Interavia*, March.

Tomasello, F. (1994), "Italian CNS/ADS initiatives", *Working Paper presented at the Third Meeting of the Automatic Dependent Surveillance Panel (ADSP/3-WP/38),* International Civil Aviation Organization, Montreal.

Walker, D. (1993), "Atlantic trials demonstrate operational capability of data link communications", *ICAO Journal*, vol. 48, no. 2, March.

Warwick, G. (1995), "GPS giant awakens", *Flight International*, 11–17January.

Westermark, H. (1994), "A candidate data link concept to support GNSS augmentation", *Working Paper presented at the First Meeting of the Global navigation Satellite System Panel (GNSSP/1-WP/28),* International Civil Aviation Organization, Montreal.

Yee, N.H.Y (1994), "Fiji's global positioning system (GPS) implementation programme", *Working Paper presented at the First Meeting of the Global Navigation Satellite System Panel (GNSSP/1-WP/5),* International Civil Aviation Organization, Montreal.

Appendix A

Annexes to the Convention on International Civil Aviation

Annexes to the Convention on International Civil Aviation

Annex 1 - Personnel Licensing

Annex 1 to the Convention on International Civil Aviation provides Standards and Recommended Practices for the licensing of flight crew members (pilots, flight engineers), air traffic controllers and maintenance technicians.

Annex 2 - Rules of the Air

Annex 2 consists of the general rules, visual rules and instrument flight rules developed by ICAO, applying without exception over the high seas and over national territories, to the extent that they do not conflict with the rules of the State being overflown. All these rules, when complied with by all concerned, help make for safe and efficient flight.

Annex 3 - Meteorological Service for International Air Navigation

The provisions of Annex 3 contribute to the safety, efficiency and regularity of air navigation. Meteorological service provides necessary weather information to operators, flight crew members, air traffic services units, search and rescue units, airport management and others concerned in aviation.

Annex 4 - Aeronautical Charts

The safe and efficient flow of air traffic is facilitated by aeronautical charts drawn to accepted ICAO Standards. The ICAO series of aeronautical charts consists of 17 types, each intended to serve specialized purposes.

Annex 5 - Units of Measurement to be Used in Air and Ground Operations

Dimensional systems to be used in air and ground operations.

Annex 6 - Operation of Aircraft

Specifications which will ensure in similar operations throughout the world a level of safety above a prescribed minimum.

Part I - International Commercial Air Transport - Aeroplane
Part II - International General Aviation - Aeroplane
Part III - International Operations - Helicopters.

Annex 7 - Aircraft Nationality and Registration Marks

Requirements for registration and identification of aircraft.

Annex 8 - Airworthiness of Aircraft

Certification and inspection of aircraft according to uniform procedures.

Annex 9 - Facilitation

Specifications for expediting the entry and departure of aircraft, people, cargo and other articles at international airports.

Annex 10 - Aeronautical Telecommunications

Standardization of communications equipment and systems (Volume I) and of communications procedures (Volume II).

Annex 11 - Air Traffic Services

Establishment and operations of air traffic control, flight information and alerting services.

Annex 12 - Search and Rescue

Organization and operation of facilities and services necessary for search and rescue.

Annex 13 - Aircraft Accident Investigation

Uniformity in the notification, investigation of and reporting on aircraft accidents.

Annex 14 - Aerodromes

Specifications for the design and operations of aerodromes (Volume I) and helicopters (Volume II).

Annex 15 - Aeronautical Information Services

Methods for the collection and dissemination of aeronautical information required for flight operations.

Annex 16 - Environmental Protection

Specifications for aircraft noise certification, noise monitoring, and noise exposure units for land-use planning (Volume I) and aircraft engine emissions (Volume II).

Annex 17 - Security - Safeguarding International Civil Aviation Against Acts of Unlawful Interference

Specifications for safeguarding international civil aviation against acts of unlawful interference.

Annex 18 - The Safe Transport of Dangerous Goods by Air

Specifications for the labelling, packing and shipping of dangerous cargo.

Appendix B

Invitation of the United States of America to the conference

INVITATION OF THE UNITED STATES OF AMERICA TO THE CONFERENCE

On September 11 the Government of the United States sent the following invitation to the Governments and Authorities listed below:

"The Government of the United States has concluded bilateral exploratory conversations with a number of other governments which have displayed a special interest on the subject of post-war civil aviation, with particular emphasis on the development of international air transport.

These discussions have indicated a substantial measure of agreement on such topics as the right of transit and non-traffic stops, the non-exclusivity of international operating rights, the application of cabotage to air traffic, the control of rates and competitive practices, the gradual curtailment of subsidies, the need for uniform operating and safety standards and the standardization of coordination of air navigation aids and communications facilities, the use of airports and facilities on a non-discriminatory basis, and the operation of airports and facilities in certain areas. It was also generally conceded that international collaboration, probably by means of an international aeronautical body, would be desirable in achieving and implementing the aforementioned objectives, although there was some diversity of opinion as to the extent of regulatory powers on economic matters which should be delegated to this international body.

The approaching defeat of Germany, and the consequent liberation of great parts of Europe and Africa from military interruption of traffic, sets up the urgent need for establishing an international civil air service pattern on a provisional basis at least, so that all important trade and population areas of the world may obtain the benefits of air transportation as soon as possible, and so that the restorative processes of prompt communication may be available to assist in returning great areas to processes of peace.

The Government of the United States believes that an international civil aviation conference might profitably be convened within the near future, for the purpose of agreeing on an increase in existing services and on the early establishment of international air routes and services for operation in and to areas now freed from danger of military interruption, such arrangements to continue during a transitional period. This conference might also agree so far as possible upon the principles of a permanent international structure of civil aviation and air transport, and might set up appropriate interim committees to prepare definitive proposals. Definitive action on such proposals, based on practical experience gained during the interim period, might be taken either as a result of a later conference, or by direct approval of the governments without the necessity of conference.

This Government suggests that the international conference proposed for the immediate future could have the following objectives:

I. (a) The establishment of provisional world route arrangements by general agreement to be reached at the Conference. These arrangements would form the basis for the prompt establishment of international air transport services by the appropriate countries.

(b) The countries participating in the conference would also be asked to agree to grant the landing and transit rights necessary for establishing the provisional route arrangements and air services referred to above.

(It would be highly desirable if each delegation were sufficiently familiar with its country's plans for international air services to permit formulation of an international air transport pattern referred to in paragraphs (a) and (b) above.)

II. The establishment of an Interim Council to act as the clearing house and advisory agency during the transitional period. It would receive and consider recommendations from each of the working committees referred to in item III; it would report upon desirable revisions in routes and services during the interim period, subject to the approval of the countries served by these routes and services; it would maintain liaison with each of the participating countries; it would supervise studies and submit information to the interested governments concerning the development of air transport during the transitional period; and would make recommendations to be considered at any subsequent international conference.

III. Agreement upon the principles to be followed in setting up a permanent international aeronautical body, and a multilateral aviation convention dealing with the fields of air transport, air navigation and aviation technical subjects; and, for the purpose of developing the details and making proposals for carrying into effect the principles so agreed, the establishment of the following working committees, which would be under the supervision of the Interim Council:

(a) A committee to follow developments relating to the establishment of the routes and services to be established under item I, to correlate traffic data, to study related problems and to recommend desirable revisions in routes and services. This committee would also make studies and recommendations concerning the future pattern of these routes and services.

(b) A central technical committee, with subordinate sub-committees, which would work closely with the committee described in subparagraph (c) below, to consider the whole field of technical matters including standards, procedures, and minimum requirements and to make recommendations for their application and adoption at the earliest practicable time.

(c) A committee to draft a proposal with respect to the constitution of a permanent international aeronautical body and a new multilateral aviation convention.

Having in mind the foregoing considerations as a basis for discussion, the Government of the United States extends a cordial invitation to your Government to participate in an international conference along the above lines, to take place in the United States beginning November 1, 1944; and in view of the time element would appreciate receiving an early response as to whether your Government can arrange to have a delegation at such conference.

This invitation is being extended to the following governments and authorities:

(a) all members of the United Nations;
(b) nations associated with the United Nations in this war;
(c) the European and Asiatic neutral nations, in view of their close relationship to the expansion of air transport which may be expected along with the liberation of Europe.

The Danish Minister and Thai Minister in Washington will be invited to attend in their personal capacities."

Appendix C

Terms of Reference of FANS I and FANS II

TERMS OF REFERENCE OF FANS I

FOREWORD

On 25 and 28 November 1983, the Council established the Special Committee on Future Air Navigation Systems (FANS) with the following terms of reference:

> To study technical, operational, institutional and economic questions, including cost/benefit effects, relating to future potential air navigation systems; to identify and assess new concepts and new technology, including satellite technology, which may have future benefits for the development of international civil aviation including the likely implications they would have for users and providers of such systems; and to make recommendations thereon for an overall long-term projection for the coordinated evolutionary development of air navigation for international civil aviation over a period of the order of twenty-five years. The committee should avoid duplication of effort and its work should be coordinated with that of existing groups in the Organization.

TERMS OF REFERENCE OF FANS II

FOREWORD

On 6 July 1989, the Council established the Special Committee for the Monitoring and Coordination of Development and Transition Planning for the Future Air Navigation System (FANS Phase II) with the following terms of reference:

The terms of reference for the Special Committee for the Monitoring and Coordination of Development and Transition Planning for the Future Air Navigation System (FANS Phase II) are:

1. To identify and make recommendations for acceptable institutional arrangements, including funding, ownership and management issues for the global future air navigation system.

2. To develop a global coordinated plan, with appropriate guidelines for transition, including the necessary recommendations to ensure the progressive and orderly implementation of the ICAO global, future air navigation system in a timely and cost beneficial manner.

3. To monitor the nature and direction of research and development programmes, trials and demonstrations in CNS and ATM so as to ensure their coordinated integration and harmonization.

Appendix D

Recommendations of the Tenth Air Navigation Conference

Recommendations of the Tenth Air Navigation Conference

Recommendation 3/1 - Development of SARPs for the ATN and associated air-ground data link ATM applications

That ICAO, as a matter of urgency, continue the development of Standards and Recommended Practices (SARPs) for both the aeronautical telecommunications network (ATN) and associated data link ATM applications to ensure that the data communication features within the CNS/ATM systems can be utilized in a timely manner.

Recommendation 3/2 - Development of SARPs for aeronautical data links

That ICAO ensure that SARPs for all forms of air-ground data links, whether terrestrial - or satellite-based, are completed in a timely manner to enable States to implement the appropriate systems for use in airspace of their responsibility.

Recommendation 3/3 - Resolution of the GNSS integrity monitoring systems

That ICAO, as a matter of urgency, continue the on-going work related to establishing the relative merit and cost of RAIM and GIC as well as determining their suitability to satisfy the requirements for monitoring GNSS integrity in order to permit an early utilization of available satellite navigation systems.

Recommendation 3/4 - Additional ATM applications of the CNS system elements

That ICAO, in the further refinement of the FANS concept and in the development of SARPs and guidance material on CNS/ATM systems elements give due recognition to additional ATM applications of the CNS system elements.

Recommendation 3/5 - Human factors studies related to the future FANS CNS/ATM systems

That work conducted by ICAO in the field of human factors pursuant to ICAO Assembly Resolution A26-9 include, *inter alia*, studies related to the use and transition to the future CNS/ATM systems and that ICAO encourage States to undertake such studies.

Recommendation 3/6 - Research and development, trials and demonstrations

That States and international organizations continue to inform ICAO on the nature and direction of their current and planned research and development efforts and trials and demonstrations in order to allow ICAO to monitor, coordinate and harmonize these activities and make the information available to States and international organizations.

Recommendation 4/1 - Institutional and legal aspects of the future air navigation systems

That further work on the institutional and legal aspects of the future air navigation systems by ICAO bodies should take account of relevant ICAO Assembly Resolutions and, *inter alia*:

a) the guiding principles proposed in Appendix D to the report on Agenda Item 4; and

b) the guidelines proposed in Appendix E to the report on Agenda Item 4.

Recommendation 4/2 - Activities related to institutional and legal aspects of the future air navigation system

That:

a) ICAO should expedite the work of the Legal Committee on the subjects of "institutional and legal aspects of the future air navigation systems", to the extent necessary to achieve implementation of the

FANS concept and the "legal aspects of the global air-ground communications"; and

b) all Contracting States should actively prepare for and participate in the studies of the Legal Committee. It is considered desirable that the Legal Committee should complete its work in time for consideration by the 29th Assembly.

Recommendation 4/3 - Required navigation performance (RNP) criteria

That ICAO, as a matter of urgency, continue the on-going work to develop the global RNP criteria, in support of early utilization of available global navigation satellite systems (GNSS).

Recommendation 4/4 - Institutional arrangements for global navigation satellite systems (GNSS)

That ICAO, as a matter of urgency, develop the institutional arrangements (including integrity aspects) as a basis for the continued availability of GNSS for civil aviation.

Recommendation 4/5 - Establishment of ICAO mechanism

That ICAO, as a matter of urgency, establish a mechanism to:

a) coordinate and monitor the implementation of the FANS concept on a global basis; and

b) provide assistance to States as required with regard to such technical, financial, managerial and legal institutional and cooperative aspects this may involve.

Recommendation 5/1 - Improvement of air traffic management

That ICAO, as a matter of urgency, develop the necessary SARPs, procedures and guidance material relating to the improvement of air traffic management in order to accommodate the anticipated growth of air traffic.

Recommendation 6/1 - Cost-effectiveness and/or cost/benefit analyses at the State level

That States:

a) individually perform their own cost-effectiveness and/or cost/benefit analyses to determine how they would be affected by the implementation of the FANS concept;

b) in such analyses give attention to the cost of training, and indirect costs and benefits, including in particular the costs of transition; and

c) inform ICAO of the methodology applied, assumptions and results of these analyses.

Recommendation 6/2 - Assistance by ICAO in the area of cost-effectiveness and/or cost/benefit analyses

That ICAO provide information and guidance to States to assist them in carrying out cost-effectiveness and/or cost/benefit analyses on the implementation of the FANS concept on a State and region-wide basis.

Recommendation 7/1 - Global transition planning

That ICAO:

a) complete and maintain a global coordinated transition plan with a view to facilitating the harmonious and timely development of global future communications, navigation, and surveillance (CNS) and air traffic management (ATM) systems; and

b) include the subject of global transition planning in its programme of regional seminars and workshops.

Recommendation 8/1 - Regional planning for implementation of the future communications, navigation, and surveillance (CNS)/air traffic management (ATM) systems

That ICAO:

a) accomplish the planning for implementation of the future CNS/ATM systems through the ICAO regional planning and implementation groups;

b) in regions where such groups have not been formed, consider, on a priority basis, forming a regional planning group;

c) ensure that the terms of reference and working arrangements for regional planning and implementation groups adequately take into account their duties and responsibilities for the planning and implementation of future CNS/ATM systems in the respective region with adequate priority; and

d) review all regional air navigation plans, including their format, as required, in the light of the global coordinated plan and the regional strategies for the implementation of future CNS/ATM system elements.

Recommendation 8/2 - Implementation planning by States

That States, in cooperation with other States as necessary, formulate plans, with a view to achieving timely implementation of the future CNS/ATM systems in conformity with planning at the regional level.

Recommendation 8/3 - ICAO's role in global implementation of the future CNS/ATM systems

That ICAO consider the establishment of a small special project team within ICAO to facilitate global coordination and harmonization of transition and implementation of the future CNS/ATM systems on a global basis.

Recommendation 8/4 - Regional participation in tests and trials

That, to facilitate implementation of the future CNS/ATM systems on a world-wide scale, ICAO encourage that the tests and trials currently under way in some regions be carried out in the other ICAO regions.

Recommendation 8/5 - ICAO assistance to facilitate the implementation of the CNS/ATM systems

That ICAO, within the framework of technical cooperation, consider the provision of assistance necessary for the tests, trials and development to facilitate the implementation of the CNS/ATM systems, particularly in developing States.

Recommendation 9/1 - Endorsement of the global CNS/ATM systems concept

That ICAO:

a) endorse the global CNS/ATM systems concept as described in paragraph 3.2 of Appendix A to the report on Agenda Item 2, and Appendix A to the report on Agenda Item 5; and

b) recognize that, in addition to the aspects referenced above, there are technical, financial, managerial and legal institutional and cooperative aspects, many of which are the subject of specific recommendations in this report, which should continue to be examined in depth by the appropriate ICAO bodies with the assistance of States and the cooperation of international organizations to benefit international civil aviation in all States.

Appendix E

Statement of ICAO policy on CNS/ATM systems implementation and operation

Statement of ICAO policy on CNS/ATM systems implementation and operation

Approved by Council (C 141/13) on 9 March 1994

In continuing to fulfil its mandate under Article 44 of the *Convention on International Aviation* by, *inter alia*, developing the principles and techniques of international air navigation and fostering the planning and development of international air transport so as to ensure the safe and orderly growth of international civil aviation throughout the world, the International Civil Aviation Organization (ICAO), recognizing the limitations of the present terrestrial-based system, developed the ICAO communications, navigation and surveillance/air traffic management (CNS/ATM) systems concept, utilizing satellite technology. ICAO considers an early introduction of the new systems to be in the interest of healthy growth of international civil aviation.

The implementation and operation of the new CNS/ATM systems shall adhere to the following precepts:

1. Universal accessibility

The principle of universal accessibility without discrimination shall govern the provision of all air navigation services provided by way of the CNS/ATM systems.

2. Sovereignty, authority and responsibility of Contracting States

Implementation and operation of CNS/ATM systems which States have undertaken to provide in accordance with Article 28 of the Convention shall neither infringe nor impose restrictions upon States' sovereignty, authority or responsibility in the control of air navigation and the promulgation and enforcement of safety regulations. States' authority shall be preserved in the coordination and control of communications and in the augmentation, as necessary, of satellite navigation services.

3. Responsibility and role of ICAO

In accordance with Article 37 of the Convention, ICAO shall continue to discharge the responsibility for the adoption and amendment of Standards, Recommended Practices and Procedures governing the CNS/ATM systems. In order to secure the highest practicable degree of uniformity in all matters concerned with the safety, regularity and efficiency of air navigation, ICAO shall coordinate and monitor the implementation of the CNS/ATM systems on a global basis, in accordance with ICAO's regional air navigation plans and global coordinated CNS/ATM systems plan. In addition, ICAO shall facilitate the provision of assistance to States with regard to the technical, financial, managerial, legal and cooperative aspects of implementation. ICAO's role in the coordination and use of frequency spectrum in respect of communications and navigation in support of international civil aviation shall continue to be recognized.

4. Technical cooperation

In the interest of globally coordinated, harmonious implementation and early realization of benefits to States, users and providers, ICAO recognizes the need for technical cooperation in the implementation and efficient operation of CNS/ATM systems. Towards this end, ICAO shall play its central role in coordinating technical cooperation arrangements for CNS/ATM systems implementation. ICAO also invites States in a position to do so to provide assistance with respect to technical, financial, managerial, legal and cooperative aspects of implementation.

5. Institutional arrangements and implementation

The CNS/ATM systems shall, as far as practicable, make optimum use of existing organizational structure, modified if necessary, and shall be operated in accordance with existing institutional arrangements and legal regulations. In the implementation of CNS/ATM systems, advantage shall be taken, where appropriate, of rationalization, integration and harmonization of systems. Implementation should be sufficiently flexible to accommodate existing and future services in an evolutionary manner. It is recognized that a globally coordinated implementation, with full involvement of States, users and service providers through, *inter alia*, regional air navigation planning and implementation groups, is the key to the realization of full benefits from the CNS/ATM systems. The associated institutional arrangements shall not inhibit competition among service providers complying with relevant ICAO Standards, Recommended Practices and Procedures.

6. Global navigation satellite system

The global navigation satellite system (GNSS) should be implemented as an evolutionary progression from existing global navigation satellite systems, including the United States' global positioning system (GPS) and the Russian Federation's global orbiting navigation satellite system (GLONASS), towards an integrated GNSS over which Contracting States exercise a sufficient level of control on aspects related to its use by civil aviation. ICAO shall continue to explore, in consultation with Contracting States, airspace users and service providers, the feasibility of achicving a civil internationally controlled GNSS.

7. Airspace organization and utilization

The airspace shall be organized so as to provide for efficiency of service. CNS/ATM systems shall be implemented so as to overcome the limitations of the current systems and to cater for evolving global air traffic demand and user requirements for efficiency and economy while maintaining or improving the existing levels of safety. While no changes to the current flight information region organization are required for implementation of the CNS/ATM systems, States may achieve further efficiency and economy through consolidation of facilities and services.

8. Continuity and quality of service

Continuous availability of service from the CNS/ATM systems, including effective arrangements to minimize the operational impact of unavoidable system malfunctions or failure and achieve expeditious service recovery, shall be assured. Quality of system service shall comply with ICAO Standards of system integrity and be accorded the required priority, security and protection from interference.

9. Cost recovery

In order to achieve a reasonable cost allocation between all users, any recovery of costs incurred in the provision of CNS/ATM services shall be in accordance with Article 15 of the Convention and shall be based on the principles set forth in the *Statements by the Council to Contracting States on Charges for Airports and Air Navigation Services* (Doc 9082), including the principle that it shall neither inhibit nor discourage the use of the satellite-based safety services.

Appendix F

ICAO state letter on policies concerning planning and implementation of CNS/ATM systems

ICAO STATE LETTER ON POLICIES CONCERNING PLANNING AND IMPLEMENTATION OF CNS/ATM SYSTEMS

Sir,

1. I have the honour to refer to the actions taken by the Council, at the second meeting of its 141st Session, on recommendations contained in the *Report of the Fourth Meeting of the Special Committee for the Monitoring and Coordination of Development and Transition Planning for the Future Air Navigation System (FANS Phase II)* (Doc 9623).

2. As a consequence of the actions taken by the Council on the aforementioned recommendations, I wish to bring to your attention the following points concerning planning and implementation of communications, navigation, and surveillance/air traffic management (CNS/ATM) systems.

3. States and international organizations which are in a position to do so are encouraged to:

a) develop and employ common procedures and specifications, where appropriate, to enable trials and demonstrations to be mounted as early as possible;

b) participate in the validation process for Standards and Recommended Practices (SARPs) of new CNS/ATM systems;

c) direct research, development, trials and demonstrations (RDT&D) at the effective transition towards the CNS/ATM systems;

d) develop programmes to extend trials and demonstrations to regions of developing States;

e) develop the necessary programmes to permit the transfer of CNS expertise, equipment, trials data and other information to developing States; and

f) utilize existing technological advances to achieve early benefits until proper CNS/ATM systems become widely available. Some examples of such existing technologies are:

1) aeronautical mobile-satellite service (AMSS) through existing available satellites;

2) automatic dependent surveillance (ADS) through existing communication media, such as very high frequency (VHF) and satellite communications;

3) global navigation satellite system (GNSS)-type services through use of global positioning system (GPS), global orbiting navigation satellite system (GLONASS), overlays, ground augmentations, etc.; and

4) aeronautical telecommunication network (ATN)-type services through, for example, use of ARINC Specification 622.

g) continue to review, address and resolve any outstanding interference issues related to interference to satellite-based aeronautical systems and report their results to ICAO.

4. Concerning the point made in 3 f) 3) above, I wish to refer you to State letter LE 4/49.1-94/89, dated 13 December 1994, which addresses the issue of ICAO's acceptance of the offer of use of the standard positioning service (SPS) of the GPS.

5. With regard to the subject contained in subparagraph 3 f) 4) above, I also wish to inform you that the Air Navigation Commission, when reviewing, at the eleventh meeting of its 137th Session, the report of the first meeting of the Aeronautical Telecommunication Network Panel (Montreal, 8 to 21 June 1994), agreed that the following points be brought to the attention of States, regional planning and implementation groups, and international organizations:

a) early use of ARINC Specification 622, although not fully compliant with the ATN, may serve as a valuable step towards the early introduction of ATM applications. States and operators should balance the potential safety and economic benefits of gaining ATM implementation experience and cost savings against the knowledge that systems based on ARINC Specification 622 will not deliver the full promise of ATN. States choosing to implement ATM applications using ARINC Specification 622 should review that specification and its performance to ensure minimum functionality to support those ATM applications;

b) transition from the systems based on ARINC Specification 622 to the ATN would involve implementation of new bit-oriented networks as well as new hardware in aircraft and on the ground. It is likely that application software and operational procedures, developed based on ARINC Specification 622, will require some modification to become ATN compliant; and

c) to assist and facilitate early development of the ATN, States and international organizations utilizing existing aeronautical telecommunication data links such as ARINC Specification 622 are requested to advise ICAO of the results of relevant-operating experiences.

6. You may wish to consider the points contained in this letter in your planning for transition to CNS/ATM systems.

Accept, Sir, the assurances of my highest consideration.

Philippe Rochat
Secretary General

Appendix G

GPS and GLONASS technical characteristics

GPS TECHNICAL CHARACTERISTICS

Technical characteristics

Satellites	24 operational satellites; 12 hour circular orbits (26 000 km radius); 55° inclinations; 6 orbit planes		
Ground control	Five globally dispersed monitor stations (down-link) Three globally dispersed ground antennas (uplink) One master control station		
Number of users	Unlimited		
Spectrum	Link 1 : L1		1 575.42 MHz
	C/A code		1.023 Mbits/s
	P-code		10.23 Mbits/s
	Navigation message		50 bits/s
	Link 2 : L2		1 227.6 MHz
	P-code		10.23 Mbits/s (current policy)
	Navigation message		50 bits/s
Method of position fixing	One-way ranging Passive user		
System outputs and accuracies		PPS	SPS
	Horizontal position	18 m (95%)	100 m (95%)
	Vertical position	28 m (95%)	157 m (95%)

	Velocity	0.2 m/s per axis not specified
	Time	180 ns (95%) 385 ns (95 %)
Acquisition time	1 to 5 minutes depending on user equipment, with stored almanac	
	Cold start: approximately 20 minutes	
Coverage	Global	
Integrity	Error detection in satellites and control system, reaction time normally less than 90 minutes (some satellites may remain out of sight of a monitor station for up to 2 hours).	
Time-scale for operational ("Block 2") implementation	Global 3D: 1992	

GLONASS TECHNICAL CHARACTERISTICS

Technical characteristics

Satellites	Twenty-four, including three standby satellites; circular orbits; rotation period eleven hours fifteen minutes; altitude 19 000 km; inclination towards the equatorial plane 64.8°; three orbital planes
Number of users	Unlimited
Frequency band	(1 602.5625 to 1 615.5) $\pm$ 0.5 MHz
Method of position fixing	Pseudo-range and radial pseudo-speed finding, the user in passive mode
System outputs and accuracies	Plane coordinates 100 m (95 per cent) Altitude 150 m (95 per cent) Speed vector components 15 cm/s (95 per cent) Time 1 μs
Signal detection time	Signal detection time depends largely on the user specific equipment performance. The satellite transmit information for navigation purposes during 30 seconds and satellites' condition information during 2.5 minutes.
Coverage	Global
Integrity	A message transmitted to the user from each satellite would contain data on troubles concerning that satellite as soon as they occur. Such information would appear in the contents of a navigation message of all satellites not later than 16 hours after the trouble occurred.

Implementation schedule	Approximately	1989-1990 - ten to twelve satellites
		1991-1995 - twenty-four satellites
Applicability to communication	The system would not be used for retransmissions of any signals or additional messages	
System upgrading	The system accuracy can be significantly increased when user operation is in a differential mode.	
	User radio link performance using an isotropic antenna	

Satellite signal effective isotropic radiated power:

- along the axis of the transmitting antenna	25 dBW
- within angles ± 15°	27 dBW
- direction of transmitting antenna polarization rotation	right-hand
- received signal power (P_s)	-(156-161) dBW
- radio link energetic potential (P_s/N)	(39-44) dBHz
- information data transmission rate	50 bits/s
- signal-to-noise ratio in the symbol (E_B/N_s)	(22-27) dB

Appendix H

Letter from the administrator of the Federal Aviation Adminstration of the United States offering the use of the global positioning system (GPS) to the international community

and

the response by the President of the ICAO Council accepting the offer by the United States

LETTER FROM THE ADMINISTRATOR OF THE FEDERAL AVIATION ADMINSTRATION OF THE UNITED STATES OFFERING THE USE OF THE GLOBAL POSITIONING SYSTEM (GPS) TO THE INTERNATIONAL COMMUNITY AND THE RESPONSE BY THE PRESIDENT OF THE ICAO COUNCIL ACCEPTING THE OFFER BY THE UNITED STATES

14 October 1994

Dr. Assad Kotaite
President of the Council
ICAO

Dear Dr. Kotaite:

I would like to commend, on behalf of the United States, the Committees on Future Air Navigation Systems (FANS) of the International Civil Aviation Organization (ICAO) for pioneering progress in the development of global satellite navigation for civil aviation. I note in this regard that the ICAO Council, on December 11, 1991, requested the Secretary General of ICAO to initiate an agreement between ICAO and Global Navigation Satellite System (GNSS) provider states concerning the duration and quality of the future GNSS.

I would like to take this opportunity to reiterate my Government's offer of the Standard Positioning Service (SPS) of the United States Global Positioning System (GPS) for use by the international community. As the United States made clear at the ICAO Tenth Air Navigation Conference and the 29th ICAO Assembly, the United States intends, subject to the availability of funds as required by United States law, to make GPS-SPS available for the foreseeable future, on a continuous, worldwide basis and free of direct user fees. This offer satisfies ICAO requirements for minimum duration of service (10 years) and freedom from direct charges. This service, which will be available as provided in the United States Government's technical sections of the Federal Radio Navigation Plan on a nondiscriminatory basis to all users of civil aviation, will provide horizontal accuracies of 100 meters (95 percent probability) and 300 meters (99.99 percent probability). The United States shall take all necessary measures to maintain the integrity and reliability of the service and expects that it will be able to provide at least six years notice prior to termination of GPS operations or elimination of the GPS-SPS.

The GPS/SPS is a candidate component of the future GNSS as envisioned by FANS. The United States believes that making the GPS available to the international community will enable states to develop a more complete understanding of this valuable technology as a component of the GNSS. The availability of GPS-SPS, of course, is not intended in any way to limit the rights of any state to control the operations of aircraft and enforce safety regulations within its sovereign airspace.

In the coming years, the international community must decide how to implement an international civil global navigation system based on satellite technology. The United States pledges its full cooperation in that endeavour and in working with ICAO to establish appropriate standards and recommended practices (SARP) in accordance with Article 37 of the Convention of International Civil Aviation (Chicago Convention). Consistent with this goal, the United States expects that SARPs developed by ICAO will be compatible with GPS operations and vice versa and that states will be free to augment GPS-SPS in accordance with appropriate SARPs. The United States will also undertake a continuing exchange of information with ICAO regarding the operation of the GPS to assist the ICAO Council in carrying out its responsibilities under the Chicago Convention.

I would be grateful if you could confirm that the International Civil Aviation Organization is satisfied with the foregoing, which I submit in lieu of an agreement. In that event this letter and your reply will comprise mutual understandings regarding the Global Positioning System between the Government of the United States of America and the International Civil Aviation Organization.

David R. Hinson
Administrator

At the 12th Meeting of its 143rd Session on 26 October 1994, the Council of ICAO considered the offer contained in your letter, and I am pleased to inform you that the arrangements outlined in the offer are acceptable to the International Civil Aviation Organization. This offer will be communicated to all Contracting States.

Accept, Sir, the assurances of my highest consideration.

Assad Kotaite

PART E

LIST OF ABBREVIATIONS

List of abbreviations

A

AAIM	Aircraft Autonomous Integrity Monitoring (GPS)
ACARS	Aircraft Communications Addressing and Reporting System
ACAS	Airborne Collision Avoidance System
ACC	Area Control Centre
ADLP	Airborne Data Link Processor
ADS	Automatic Dependent Surveillance
ADS-B	Automatic Dependent Surveillance - Broadcast
AEEC	Airlines Electronic Engineering Committee
AES	Aircraft Earth Station
AFS	Aeronautical Fixed Service
AFTN	Aeronautical Fixed Telecommunication Network
AGDLS	Air-Ground Data Link System
AIDC	ATC Interfacility Data Communications
AMS(R)S	Aeronautical Mobile Satellite (Route) Service
ANC	Air Navigation Commission
ANP	Air Navigation Plan
AOC	Aeronautical Operational Control
AOP	Aerodrome Operations
AOR-E/W	Atlantic Ocean Region East/West (Inmarsat)
APANPIRG	Asia/Pacific Air Navigation Planning and Implementation Regional Group
APIRG	AFI Planning and Implementation Regional Group

ARINC	Aeronautical Radio Inc.
ASECNA	Agency for the Safety of Aerial Navigation in Africa and Madagascar
ASM	Airspace Management
ATA	Air Transport Association of America
ATAG	Air Transport Action Group
ATFM	Air Traffic Flow Management
ATM	Air Traffic Management
ATN	Aeronautical Telecommunication Network
AWO	All Weather Operations
ATS	AIR TRAFFIC SERVICES

C

CIS	Co-operative Independent Surveillance
CPDLC	Controller/Pilot Data Link Communications

D

DARPS	Dynamic Air Route Planning Study
DGNSS	Differential Global Navigation Satellite System
DGPS	Differential Global Positioning System
DME	Distance Measuring Equipment
DOTS	Dynamic Ocean Track System (FAA)

E

EANPG	European Air Navigation Planning Group
EASIE	Enhanced ATM and Mode S Implementation in Europe
EATCHIP	European Air Traffic Control Harmonization and Integration Programme
EATMS	European Air Traffic Management System
ECAC	European Civil Aviation Conference
EGNOS	European Global Navigation Overlay System
EUROCAE	European Organization for Civil Aviation Equipment
EUROCONTROL	European Organization for the Safety of Air Navigation

F

FAA	Federal Aviation Administration
FANS	Future Air Navigation Systems
FANS	Special Committee on Future Air Navigation Systems
FDMA	Frequency Division Multiple Access
FDPS	Flight Data Processing System
FIR	Flight Information Region
FMS	Flight Management System

G

GAIT	Ground-Based Augmentation and Integrity Technique
GATE	EANPG Working Group for Air Traffic Management in the Eastern Part of the ICAO EUR Region
GEO	Geostationary Earth Orbit
GES	Ground Earth Station
GIB	GNSS Integrity Broadcast
GIC	GNSS Integrity Channel
GLONASS	Global Orbiting Navigation Satellite System
GNSS	Global Navigation Satellite Systems
GPS	Global Positioning System
GREPECAS	CAR/SAM Regional Planning and Implementation Group

H

HF	High Frequency

I

IACA	International Air Carrier Association
IAOPA	International Council of Aircraft Owner and Pilot Associations
IATA	International Air Transport Association
IBAC	International Business Aviation Council

ICAO	International Civil Aviation Organization
ICCAI	International Coordinating Council of Aerospace Industries Associations
IFALPA	International Federation of Air Line Pilots' Associations
IFATCA	International Federation of Air Traffic Controllers' Associations
IFR	Instrument Flight Rules
ILS	Instrument Landing System
IMC	Instrument Meteorological Conditions
Inmarsat	International Maritime Satellite Organization
INS	Inertial Navigation System
IPACG	Informal Pacific Ocean ATS Coordinating Group
IRS	Inertial Reference System
ISDN	Integrated Services Digital Network
ISPACG	Informal South Pacific ATS Coordination Group

K

KLADGNSS	Kinematic Local Area Differential GNSS

L

LAAS	Local Area Augmentation System
LADGNSS	Local Area Differential GNSS
Loran	Long Range Air Navigation System

M

MASPS	Minimum Aircraft System Performance Specification
MIDANPIRG	Middle East Air Navigation Planning and Implementation Regional Group
MLS	Microwave Landing System
MMALS	Multimode Approach and Landing System
MMR	Multimode Receiver
MOPS	Minimum Operational Performance Standards

MTSAT	Multi-Functional Transport Satellite

N

NAT SPG	North Atlantic System Planning Group
NDB	Non-Directional Radio Beacon
NEAN	North European ADS-B Network
NSE	Navigation System Error

O

OLDI	On-Line Data Interchange
OPS	Operations
OR	Operational Requirements
OSI	Open System Interconnection

P

PANS	Procedures for Air Navigation Services
PET	Pacific Engineering Trial

R

RAC	Rules of the Air and Air Traffic Services
RACGAT	Russian/American Coordinating Group on Air Traffic Control
RAIM	Receiver Autonomous Integrity Monitoring (GPS receiver)
RAN	Regional Air Navigation (RAN) Meeting
R&D	Research and Development
RDT&D	Research, Development, Trials and Demonstrations
RGIC	Ranging GNSS Integrity Channel
RNAV	Area Navigation
RNP	Required Navigation Performance
RNPC	Required Navigation Performance Capability

RPG	Regional Planning Group
RTCA	RTCA
RTD	Research and Technical Development

S

SARPs	Standards and Recommended Practices
SATCOM	Satellite Communication
SITA	Societe Internationale de Telecommunications Aeronautiques
SMGCS	Surface Movement Guidance and Control Systems
SOIR	Simultaneous operations on Parallel or Near-Parallel Instrument Runways
SSR	Secondary Surveillance Radar
STDMA	Self-Organizing Time Division Multiplex

T

TARTAR	Meetings for the planning and coordination of implementation of ATS Routes Through the Airspace of the Russian Federation, including Middle Asia
TCAS	Traffic Alert and Collision Avoidance System
TDMA	Time Division Multiple Access
TREAT	Trans EUR/ASIA Air Traffic Management Initiative (for FANS 1 equipped Aircraft
TWDL	Two-Way Data Link

V

VFR	Visual Flight Rules
VHF	Very High Frequency
VMC	Visual Meteorological Conditions
VOR	VHF Omnidirectional Radio Range

W

WAAS	Wide Area Augmentation System
WADGNSS	Wide Area Differential GNSS
WADGPS	Wide Area Differential Global Positioning System
WAFC	World Area Forecast Centre
WGS-84	World Geodetic Reference System 1984
WMC	World Meteorological Centre

Index